Endolichenic Fungi: Present and Future Trends

Manish Tripathi • Yogesh Joshi

Endolichenic Fungi: Present and Future Trends

Springer

Manish Tripathi
Department of Botany
Kumaun University
Almora, Uttarakhand, India

Yogesh Joshi
Department of Botany
Kumaun University
Almora, Uttarakhand, India

Department of Botany
University of Rajasthan
Jaipur, Rajasthan, India

ISBN 978-981-13-7267-4 ISBN 978-981-13-7268-1 (eBook)
https://doi.org/10.1007/978-981-13-7268-1

This Springer imprint is published by the registered company Springer Nature Singapore Pte Ltd.
The registered company address is: 152 Beach Road, #21-01/04 Gateway East, Singapore 189721, Singapore

In Loving Memory Of

Father of Indian Lichenology

Dr. D. D. Awasthi

(28/9/1922 – 21/8/2011)

Loved and Remembered

Foreword

Over the past decade, studies on endophytes have received increasing research attention because of their potential to produce unique secondary metabolites with promising pharmaceutical properties. Endophytes in cryptogamic plants such as algae, fungi, lichens, and bryophytes have not received adequate attention throughout the world. I am delighted to introduce an eminently readable book, *Endolichenic Fungi: Present and Future Trends*, to academicians, students, and researchers, which Dr. Manish Tripathi and Dr. Yogesh Joshi have produced. Though lichens themselves are symbiotic association of algae and fungi, mere presence of endophytic fungi as endophyte further makes them a unique and peculiar organism. These endophytes represent almost unlimited and sustainable sources of bioactive and chemically novel natural products with the potential for utilization in an array of medical, agricultural, and industrial applications. Since there are a number of books available on lichens and endophytes, not a single comprehensive book on taxonomy and biology of endolichenic fungi is published by any author across the globe. The limited understanding of such important microorganisms is a testament to the fact that the "age of discovery" is just beginning. This book not only comprises of the first full survey on what the lichens hold within – endophytic fungi – but also is first of its type across the globe. The book is well-written in the most authentic and lucid manner and is a treasure of information about recent developments in the field of endolichenology, demonstrating a wealth of interesting details. This book will surely enlighten the new minds and become a source of inspiration and information for those who wish to initiate research in this fascinating area.

The authors didn't need any sort of acclamation, but I can't resist expressing my deep admiration for their enthusiasm and relentlessly writing on a novel field of lichenology, thus opening a new horizon in this branch of science which is quite neglected not only in India but throughout the world.

I strongly feel that this timely publication is a unique and important addition to the literature on endolichenic fungi which will serve the needs of academicians, researchers, biologists, and anyone who is interested in endolichenic research. I congratulate the authors for their endeavor for bringing out this comprehensive and qualitative important contribution not only to Indian lichenology but also to world lichenology that will guide young minds for a long time to come.

D. K. Upreti
FNA, FNASc, FES, FISEB
CSIR-Emeritus Scientist

Preface

Although generally lichens are considered to be a classic case of mutualism, it is now well established that they form associations with a consortium of incredibly numerous and diverse microbes, which dwell either on their surface (lichenicolous or incidental fungi) or within the thallus (endolichenic fungi, bacteria, yeasts). This book is the first full survey on what the lichens hold within – endolichenic fungi. Interestingly, these endolichenic fungi have converged to perform functions different from what they perform in the free form in nature, possibly allowing the fungi to survive in different lifestyles to meet the challenge of variable or deleterious environmental conditions.

There are several fields where an improved understanding of endolichenic fungi is necessary. First, endolichenic fungi may be a more important part of host lichen biology than previously recognized. Endolichenic fungi and other lichen-associated fungi are often neglected when studying lichen or the effects of stressors such as air pollution and disease on lichens. However, endolichenic fungi have a pivotal role in, for example, the interaction between lichens and insect pests and can be susceptible to the same stressors as their lichen host, with possible consequences for important mutualistic interactions. These examples illustrate the necessity of taking a holistic approach to the study of lichens, including associated endolichenic fungi and other associated fungi.

Second, endolichenic fungi have great biotechnological potential. Bioactivity of endolichenic fungi is being harnessed for applications in various arenas worldwide. The fact that they provide a widely unstudied and diverse source of completely new bioactive compounds has already raised the interest of pharmaceutical industry. The bioactive compounds produced by endophytes have unlimited prospects for development as leads for various drug compounds, cosmetics, food preservatives, components for wood and paper industry, drapery, and so on. Advances are currently being made on biocontrol of insect pests by endolichenic fungi, which may bring new biotech applications on the market within the near future. Endolichenic fungi-based applications could, and should, become much more extensive and important to mankind than they are today.

Essential for promotion of all research fields is a better understanding of the evolution, diversity, and patterns of host-endolichenic fungi interactions in the wild. Such knowledge is necessary for predicting behavior of endolichenic fungi (mutualistic vs pathogenic) with respect to the host under specific conditions, or in new ecosystems.

Fortunately, new technologies and methods that are discussed throughout this book have the potential to close this knowledge gap and bring the "myco-lichenology" to a new level. Thanks to the rapid progress in sequence technology, we are now poised to explore the uncultured portion of endolichenic fungi across a large number of lichen species, tissues, geographic locations, and environmental conditions. Culture-independent approaches that allow characterization of a large number of samples at depth can partition the community in core versus transient species that may have different functional roles in regard to the host. Such information might prove useful when selecting isolates for, e.g., in-vitro assays or biocontrol studies.

Genome sequencing has the potential to validate and extend current models of endophyte entry, colonization, persistence, and beneficial host interaction. Finally, metagenomics, single-cell genomics, and expression of pathways in heterologous hosts hold great promise to discover endolichenic fungal metabolites and yield insights into the ecological role of endolichenic fungi that elude current culturing attempts.

We can expect that the continuous development of these techniques for the study of endophytes will fully reveal their biology in lichen. As endolichenic fungi are proving to be the crucial, even essential components of lichen's life, their significance for mankind might equal, or even exceed, that of other lichen symbionts.

Our goal was to organize a book that will be a major work on the endolichenic fungi, serving as the first point of reference for this important group of organisms. We envisioned producing a handbook for professional researchers and research students in the field, as well as a textbook for graduate and undergraduate courses. Although model species of lichens and endolichenic fungi are covered in detail, we strove to represent the diversity of the endolichenic fungi. The chapters in this book cover the topic in very a reader-friendly and stepwise manner from what are endolichenic fungi, how to isolate and identify them, how to isolate the secondary metabolites from them, etc. Every effort has been made to incorporate latest data from all the sources and to provide well-grounded evaluation of the literature.

Uttarakhand, India

Manish Tripathi
Yogesh Joshi

Acknowledgments

Having an idea to write a book and executing that idea into an actual book are two completely different things. As this is our first book, so the experience of writing this book is both challenging as well as rewarding.

This work would not have been possible without the financial support of the University Grants Commission, New Delhi [File No. 41-488/2012 (SR)], Council of Scientific and Industrial Research, New Delhi [File No. 38(1441)/17/EMR-II], and Integrated Eco-development Research Program of G. B. Pant National Institute of Himalayan Environment and Sustainable Development [File No. GBPI/IERP/16-17/16/175].

The authors would like to express their gratitude to **Prof. Hema Joshi** (Head, Department of Botany, S.S.J. Campus, Kumaun University, Almora, Uttarakhand), **Dr. Manju Sharma** (Head, Department of Botany, University of Rajasthan, Jaipur, Rajasthan) and **Prof. Alpana Kateja** (Principal, Maharani College, University of Rajasthan, Jaipur, Rajasthan) who not only were quite supportive but also provided laboratory facilities and academic time to carry out research on this aspect.

We take pride in acknowledging the insightful guidance of **Dr. D.K. Upreti**, Senior Principal Scientist, CSIR-National Botanical Research Institute, Lucknow, for sparing his valuable time whenever we approached him and showing us the way ahead. His support, encouragement, and credible ideas have helped us in completing this book.

We wish to express our appreciation to **Dr. K.P. Singh, Dr. G.P. Sinha, Prof. P.K. Divakar, Dr. Sanjeeva Nayaka, Dr. Rajesh Bajpai, Dr. T.A.M. Jagadeesh Ram, Emeritus Prof. R.C. Gupta, Prof. S.S. Gahalain, Prof. S.C. Sati, Prof. Neerja Pande, Prof. Lalit Tewari, Prof. S.S. Bargali, Prof. R.C. Dubey, Prof. N.S. Atri, Prof. T. S. Suryanarayanan, Mr. Jyoti Tandon**, and all the members of Indian Lichenological Society and Indian Mycological Society for their constant encouragement and are also grateful to all of those with whom we have had the pleasure to work during this and other related projects.

We also wish to thank anonymous reviewers for their timely remarks about the manuscript. Thanks are also extended to all the members of Springer team (**Ms. Akanksha Tyagi, Ms. Raman Shukla, Ms. Ulaganathan Padmapriya, Ms. Uma Maheswari Srinivasan**) and other staff of Springer publication house for their co-operation and valuable suggestions.

Nobody has been more important to us in the pursuit of this project than the members of our family. We would like to thank our parents, whose love and guidance are with us in whatever we pursue.

Contents

About the Authors

Manish Tripathi did his M.Sc. in Botany and received Ph.D. in 2017 on *Diversity and antimicrobial activity of endolichenic fungi isolated from some macrolichens of Kumaun Himalaya* from Kumaun University. Currently, he is working as Guest Faculty at the Department of Botany, Kumaun University, S.S.J. Campus, Almora. Prior to working on this post, he worked as Junior Research Fellow (UGC Sponsored Project) and Research Associate (CSIR Project) and has published 25 papers in national and international journals. He was also bestowed with Uttarakhand Young Scientist Award in 2013 given by Uttarakhand Congress for Science and Technology (UCOST) for his novel work in the field of endolichenic fungi. His research is focused on lichen diversity studies, systematics of endolichenic fungi, and characterization and evaluation of secondary metabolites of endolichenic fungi for their antibiotic potential.

Yogesh Joshi before joining Department of Botany, University of Rajasthan, Jaipur, in 2018 as Associate Professor, served Department of Botany, Kumaun University, S.S.J. Campus, Almora, from 2011 to 2018 on the post of Assistant Professor. For a short duration (March 2016 to June 2016), he also worked as Scientist E1 at Kerala Forest Research Institute (KFRI), Peechi, Kerala. He did his Master's and Doctorate from Kumaun University, Nainital, and Post Doctorate from Sunchon National University, Suncheon, South Korea, and has published about 100 scientific papers in national and international reputed journals. He has successfully completed many projects funded by various agencies, viz., University Grants Commission (UGC), Science and Engineering Research Board (SERB), Department of Science and Technology (DST), Council of Scientific and Industrial Research (CSIR), and G.B. Pant National Institute of Himalayan Environment and Sustainable Development (GBPNIHESD), and is recipient of many academic awards, including Prof. Y.S. Murty Gold Medal (2015), Uttarakhand Young Scientist Award (2015), Dr. P.D. Sethi Memorial National Award (2016), Iyengar-Sahni Medal (2017), etc. He is a Fellow of the Association of Plant Taxonomy (FAPT) and Life Member of various prestigious societies, viz., Indian Botanical Society, Indian Mycological Society and Indian Lichenological Society. He has 10 years of research and teaching experience, and during that period, he supervised/co-supervised eight students for their Doctorate Degree.

What are Lichenized Fungi?

Abstract Conventionally lichens are considered as an example of symbiotic involvement between a fungus and one or more algae but it is widely debated and deserves further investigation. With the discovery of endolichenic and lichenicolous fungi this relationship becomes much more complex to explain. The lichen thallus in itself is a comparatively established and proportionate symbiotic association with both heterotrophic and autotrophic participants. If the parasitic lichenicolous fungi are considered as decomposers of this ecosystem then the lichen can be considered as an autonomous minuscule biological network. The lichens produces plethora of secondary metabolites, such as, phenolic compounds, dibenzofurans, depsides, depsidones, depsones, lactones, quinones and pulvinic acid derivatives, which are accumulated externally on the hyphae rather within the cells. These compounds not only play substantial role in characterizing a species but also have bioactive potential and allow lichens to be used as food, fodder, dyes, medicines and pharamaceuticals. Besides this, since time immemorial, lichens have been used as best bio-indicators of air pollution. But now days, these tiny creatures are facing high risk of extinction and needed to be conserved. These organisms can be protected by the preservation of habitats, in-situ conservation of ecological niches, and also by promoting research on lichens. This chapter discusses the unique details about lichens and the rationale of studying lichens like other higher plants.

Theophrastus (370–285 BC) employed the term 'lichen' denoting apparent growth on the bark of olive trees. Tournefort while preparing his *Elements de Botanique* (1694) came across several quite unrelated plants that had been categorized as lichens. He delimited the group and created a genus *Lichen* to accommodate such plants in question which are having similar fruit morphology (Mitchell 2007). Antoine de Jussieu, was the first botanist who stated that fungi and lichens are related. In his paper, Jussieu (1730) remarked apropos of fungi that "if one looks in the classes of plants for a genus they resemble and to which they may be compared, the only one found is *Lichen*". However, the foremost taxonomic account under genus *Lichen* comprising 80 species was published under the 24th Class of Cryptogamie-Algae in *Species Plantarum* by Linnaeus (1753). He assigned *Lichen*

M. Tripathi, Y. Joshi, *Endolichenic Fungi: Present and Future Trends*,
https://doi.org/10.1007/978-981-13-7268-1_1

not to his order Fungi but, with a few exceptions, to the Algae. Erik Acharius, who is accredited as *Father of Lichenology*, in his authoritative *Lichenographia universalis* (1810) declared "*to summarize from a complete and careful examination of all the parts, I have concluded that lichens represent a special, natural, group separate from other cryptogamic plants*". He created various terminologies for different unique structures of lichens, described numerous novel genera and species and also established the first classification of lichen fungi, "lichens", in his monumental works *Lichenographiae Svecicae Prodromus*, *Methodus qua omnus detectos Lichenes*, *Lichenographia universalis* and *Synopsis methodica lichenum* (Acharius 1798, 1803, 1810 and 1814, respectively). The plant body of lichen i.e. thallus, had been believed to be a single plant till 1867, when Schwendener, propounded the dual hypothesis for this body, mentioning that the lichen thallus is a complex body formed of a fungus and an alga. Throughout nineteenth and twentieth century's, lichen fungi have been placed in its own group **Lichenes**, based on its symbiotic life form as articulated by its complex thalli (e.g. Henssen and Jahns 1974; Poelt 1973; Zahlbruckner 1907, 1926). They were included under fungi in International Code of Botanical Nomenclature implemented by the Thirteenth International Botanical Congress (Voss et al. 1983).

Time to time several definitions of lichens came into existence (Box 1). They are by definition not homogeneous organisms, but are symbiotic association of two utterly discrete organisms – the mycobiont and one or more photobionts (the photobiont may be the member of either cyanophyceae or chlorophyceae), in which the mycobiont is the exhabitant (Hawksworth 1988). Around 85% species of lichens

Box 1 Different Definitions of Lichens

- Simon Schwendener (1867) proposed his dual theory of lichens, that lichens are a combination of fungi with algae or cyanobacteria, whereby the true nature of the lichen association began to emerge (Honegger 2000).
- Ahmadjian and Jacobs (1983) have interpreted the relationship of the two symbionts as **controlled** or **balanced** parasitism in which the parasite (mycobiont) does not allow complete annihilation of the host (photobiont), and balance is maintained for the growth of the host and the parasite. Ahmadjian (1993) has given the evidence to suggest that the lichen symbiosis is parasitic or commensalistic, rather than mutualistic. The photosynthetic partner can exist in nature independently of the fungal partner, but not vice versa. Photobiont cells are routinely destroyed in the course of nutrient exchange. The association is able to continue because reproduction of the photobiont cells matches the rate at which they are destroyed (Ahmadjian 1993). The fungus surrounds the algal cells (Speer and Ben 1997), often enclosing them within complex fungal tissues unique to lichen

(continued)

Box 1 (continued)

associations. In many species the fungus penetrates the algal cell wall (Speer and Ben 1997), forming penetration pegs (haustoria) similar to those produced by fungi that feed on a host (pathogenic fungi) (Honegger 1988).

- Lichen is a stable self-supporting association of a mycobiont (fungus) and a photobiont (alga) in which the mycobiont is the exhabitant (Hawksworth 1988).
- According to Ahmadjian (1993) lichen may be described as "an association between a fungus, usually an ascomycete but in a few cases a basidiomycete or deuteromycete, and one or more photosynthetic partners, generally green algae or cyanobacteria.
- A lichen is "an ecologically obligate, stable mutualism between an exhabitant fungal partner (the mycobiont) and inhabitant population of extracellularly located unicellular or filamentous algal or cyanobacterial cells (the photobiont)" (Hawksworth and Honegger 1994).
- Galun and Kardish (1995) define the lichens as symbiotic associations between a fungus and a cyanobacterium (cyanolichens) or a green alga (phycolichens), joined to form a new biological entity different from its individual components. Both bionts appear in nature among a mixture of millions of nonsymbiotic microorganisms and they have to select each other for a compatible combination.
- Lichen is a composite organism that arises from algae and/or cyanobacteria living among filaments (hyphae) of a fungus in a mutually beneficial (symbiotic) relationship (Brodo et al. 2001).
- According to Armstrong (2004) lichen is an intimate association between two quite different microorganisms, viz., an alga and a fungus resulting in a substratum.
- Lichens are said to be "species", but what is meant by "species" is different from what is meant for plants, animals, and fungi, for which "species" implies a common ancestral lineage. Because lichens are combinations of members of two or even three different biological kingdoms, these components *must* have a *different* ancestral lineage from each other. By convention, lichens are still called "species" anyway, and are classified according to the species of their fungus, not the species of the algae or cyanobacteria. Lichens are given the same scientific name (binomial name) as the fungus in them, which may cause some confusion. The alga bears its own scientific name, which has no relationship to the name of the lichen or fungus (Kirk et al. 2008).
- Lichens are classified by the fungal component. They are given the same scientific name (binomial name) as the fungus species in the lichen and are being integrated into the classification schemes for fungi. The alga bears its own scientific name, which bears no relationship to that of the lichen or fungi (Kirk et al. 2008).

(continued)

Box 1 (continued)

- Jovan (2008) stated that although appearing to be a single organism, lichen is actually a symbiotic partnership between a fungus and one or more photosynthetic organisms, an alga or cyanobacterium. Typically the fungal partner provides most of the composite organism's structure and mass, thus trading physical protection for carbohydrates manufactured by the photosynthetic partner. Together the fungus and its partner(s) can inhabit a much wider variety of habitats and conditions than any could on their own.
- *"Lichens are fungi that have discovered agriculture"*.......... has been attributed to the lichenologist Trevor Goward (Sanders 2001). The lichen fungi (kingdom Fungi) cultivate partners that manufacture food by photosynthesis. Sometimes the partners are algae (kingdom Protista), other times cyanobacteria (kingdom Monera). Some enterprising fungi exploit both at once.
- According to Nash III (2008) lichens are a group of fungi with a specific nutritional strategy which consists of absorbing carbon from autotrophic algal cells. This is a little ecosystem in which a dynamic balance is created between the fungus and the alga, and any change in habitat conditions can disturb this balance.
- Cobanoglu et al. (2010) proclaimed lichens as a symbiotic association between fungi and algae, so called "lichenized fungi" including over 20,000 species all over the world.
- Lichen thallus can be regarded as a '*functional organismic community*' or as a microhabitat with a huge variety of coexisting fungal, algal and bacterial genotypes (Boonpragob et al. 2012).
- The same fungus growing in combination with different algae and/or cyanobacteria, can produce lichens that are very different in most properties, meeting non-DNA criteria for being different "species". Historically, these different combinations were classified as different species. When the fungus is identified as being the same using modern DNA methods, these apparently different species get reclassified as the *same* species under the current (2014) convention for classification by fungal component. This has led to debate about this classification convention. These apparently different "species" have their own independent evolutionary history.
- Lichens also host basidiomycete yeasts embedded in their cortex, constituting it an important part of thallus (Spribille et al. 2016).
- Lichens are relatively '*self-contained miniature ecosystems*' in and of themselves, possibly with more microorganisms living with the fungi, algae, and/or cyanobacteria, performing other functions as partners in a system that evolves as an even more complex composite organism (holobiont) (Gerson and Seaward 1977; Honegger 1991; Barreno et al. 2008; Grube et al. 2009; Casano et al. 2011).
- Lichens are a complex life form that is a symbiotic partnership of two separate organisms, a fungus and an alga. The dominant partner is the fungus, which gives the lichen the majority of its characteristics, from its thal-

(continued)

Box 1 (continued)

lus shape to its fruiting bodies. The alga can be either a green alga or blue-green alga, otherwise known as cyanobacteria. Many lichens will have both type of alga (USDA).

- Lichens consist of a symbiotic association between a photosynthetic microorganism (photobiont), often a cyanobacterium or green alga, and a fungus (mycobiont), usually an ascomycete but occasionally a basidiomycete (Galloway 1992). This highly successful symbiosis permits lichens to colonize a variety of harsh environments and even thrive in recently disturbed habitats (Medlin 1996). In addition to their role as pioneers, lichens have important ecological functions in food webs (Hale and Cole 1988), mineral and nutrient cycling (Nieboer et al. 1978), carbon fixation (Galloway 1992), and nitrogen fixation (Forman 1975).
- Lichens are not only the poster children of symbiosis (De Bary 1879), but they are also difficult to place into the predictable constructed categories that we turn to so frequently in both biology and politics. Lichens are in fact far more than their understood identity of a symbiosis between fungi and algae; there are many more components involved, including fungi, algae and bacteria, in addition to a potential third partner in the lichen symbiosis (Aschenbrenner et al. 2016; Spribille et al. 2016; Muggia and Grube 2018; Scharnagl 2019).

As per different dictionaries

- Any of numerous complex plantlike organisms made up of an alga or a cyanobacterium and a fungus growing in symbiotic association on a solid surface (such as on a rock or the bark of trees). The main body of the lichen, known as the thallus, is formed by fungal filaments which surround the photosynthetic algal or cyanobacterial cells (Merriam Webster dictionary).
- A simple slow-growing plant that typically forms a low crusty, leaf like, or branching growth on rocks, walls, and trees. Lichens are composite plants consisting of a fungus that contains photosynthetic algal cells. Their classification is based upon that of the fungal partner, which in most cases belongs to the subdivision ascomycotina, and the algal partners are either green algae or cyanobacteria (Oxford dictionary).
- Composite organism made up of a fungus, usually an ascomycete, which grows symbiotically with an alga or acyanobacterium and characteristically forms a crust like or branching growth on rocks or tree trunks (The free dictionary).
- A type of organism that consists of an alga (plant without leaves or roots and having chlorophyll) and a fungus (plant without leaves, flower, roots and chlorophyll) growing together (Cambridge dictionary).
- Lichen is a group of tiny plants that looks like moss and grows on the surface of things such as rocks, trees, and walls (Collins dictionary).

have green algae as symbionts, *ca.* 10% have cyanobacteria and less than 5% have both green algae as primary symbionts and cyanobacteria as secondary symbionts in cephalodia. The majority of the ground tissue of lichen is made up of mycobiont (90%), while photobiont contribute usually less than 10% of the total thalline mass (Awasthi 2000). Since mycobiont predominates in the lichen thallus, hence, the gross morphology of lichen thallus is generally established by mass, nature and modifications of the fungal hyphae. Repeated septation, branching and various levels of density, coalescence or conglutination in hyphae results in the formation of diverse type of tissues (e.g. paraplectenchymatous, prosoplectenchymatous, chalaroplectenchymatous, scleroplectenchymatous, scleroprosoplectenchymatous, palisade plectenchymatous), forming various structures (viz. cortex, pseudocortex, phenocortex, epicortex, lithocortex, epinecral layer), which are quite necessary for a durable and proper functioning of the symbiotic association in a lichen thallus, as these tissues protect the photobiont (except most of the cyanobacteria which have a mucilaginous scabbard protecting them from dehydration) from dehydration in a lichenized terrestrial situation (Ahmadjian 1993). On the basis of the distribution of photobiont, two distinctly different conditions of lichen thalli are known – (i) *homoiomerous*: the two symbionts are uniformly distributed all over the thallus (e.g. *Collema*), and (ii) *heteromerous*: the photobiont and mycobiont are stratified in a layer. The main subdivisions of layers are upper cortex, algal layer, medulla and lower cortex, and these layers may include various tissue types.

Lichens are fungi that have evolved a unique physiology. Instead of acquiring carbon and energy by decomposing organic matter they used to associate with living green algae or cyanobacteria, which supply them sugar produced during photosynthesis. The fungus forms the physical structure enclosing the algae and determines the morphology and anatomy of the lichen. The composite structure – called a *thallus* – is unique for each species of lichen forming fungus. The individual algal cells do not receive anything in return for supply of sugar, but at the population level they get benefited by expanding their ecological and geographical range significantly. In addition to performing photosynthesis cyanobacteria are able to fix atmospheric nitrogen, which is incorporated into amino acids and is made available to the fungus. In this way the fungus is also supplied with an efficient nitrogen source (Søchting 1999).

Although the dual nature of most lichens is now widely recognized, it is rarely acknowledged that some lichens have symbioses connecting three (tripartite lichens) or more partners. The prospective association of algal and fungal partners may reasonably be intricate, and a precise categorization of various associations was proposed by Rambold and Triebel (1992). The lichen thallus resembles neither of the synthetic partners and in many studies is implicitly treated as individual, even though it may be a symbiotic entity involving members of three kingdoms. Hence, from a genetic and evolutionary view point, they undoubtedly cannot be considered as individuals and this fact has key insinuations for various fields of exploration, such as developmental and reproductive studies (Nash 2008).

Since time immemorial, the nature of the lichen symbiosis is widely debated and deserves further investigation (Nash 2008). Reinke (1872) suggested the existence of interdependence between the two constituents for mutual benefit and called the relationship consortism or mutualism. De Bary (1879) called the relationship as symbiosis wherein both the components get benefitted though not necessarily to an equal degree. Because of the presence of dead algal cells in the lichen thallus, Elenkin (1902) suggested the relationship to be parasitism or endosaprophytism. Nienburg (1917) was of the view that since the alga in lichen has advantages as well as suffers harm, the relationship should be termed helotism. Recently, lichens are regarded as an example of controlled parasitism or balanced parasitism, because the fungus seems to obtain most of the benefits and the photobiont may grow more slowly in the lichenized state than when free-living (Ahmadjian and Jacobs 1983; Ahmadjian 1993), i.e. the parasite (mycobiont) does not allow complete annihilation of the host (photobiont), and a balance is maintained for the growth of the host and parasite. In fact, the relationship becomes much more complex, especially when additional lichenicolous fungi (Lawrey and Diederich 2003) grow on lichen thallus. These are different fungi from the dominant mycobiont, and they may have a parasitic, commensalistic, mutualistic or saprophytic/saprobic relationship with the lichen (Rambold and Triebel 1992) and the morphology and physiology of which are, as yet, little understood. Whatever it be, the lichen symbiosis has always bought fungus and alga substantial profits. The fungus receives carbohydrate from the algae necessary for its existence, while the alga is protected in the enveloping intertwining fungus from rapid water loss, from intensive solar rays, or from easy clutches of algae feeding animals (Wirth 1995). With the help of symbiosis, the involved fungus and alga in the lichen, have considerably widened their ecological potentiality and succeeded in colonizing the most harsh and inhospitable habitats on earth from the tropics to polar regions, where separately they would be rare or non-existent.

The degree of lichenization ranges from arbitrarily associated few algal cells and a fungus (e.g. some Caliciales) to the well organized thallus with all the distinct layers (medulla and cortex) made up of fungal and algal partner (Nash 2008). A single definition cannot sufficiently cover all the aspects of algal-fungal associations within lichens because of this remarkable variation in lichenization. The photobiont individually and in association with the mycobiont has the capacity to greatly influence the morphology of the lichens thallus. After the use of molecular techniques it has come in light that in few cases in nature the same mycobiont is capable of forming two entirely different interconnected thalli with a cyanobacterium and a green alga (Armaleo and Clerc 1990). These distinct morphotypes are known as photosymbiodemes, and their incidence entails the ontogenetic control by the photobiont (Nash 2008). The degree of symbiotic obligation between algal and fungal partners within lichen thallus also varies. The fact that to which extent the algal species has been documented living in both free living and lichenized states (Beck 2002) is not well established because comparatively not much lichen algae including both cyanobacteria and green algae have been identified to species level. However, most of the lichens are extremely selective in their choice of photobiont (Beck et al. 1998; Rambold et al. 1998). On the other hand, the systematics of the fungal partner is well established. Most of the fungal partners have an obligatory symbiosis towards lichenization because the isolated mycobionts grow very slowly and are questionable to survive well in the free-living state because of the struggle with other fungal species or consumption by other organisms; although the specificity of a fungal partner for a particular algal partner is not noteworthy (Friedl 1989; Ihda et al. 1993).

The lichen thallus in itself is a comparatively established and proportionate symbiotic association including both heterotrophic and autotrophic components (Nash 2008). From this viewpoint, lichens can be considered as an infinitesimal biological network (Farrar 1976; Seaward 1988), especially when lichenicolous fungi, because of their parasitic nature, considered as decomposers of this network. Along with the primary fungal and algal partner other symbiotic partners also cohabit lichen thallus, and the fungal population other then the primary fungi (i.e. the mycobiont), is termed as accessory or secondary fungi (Zopf 1897; des Abbayes 1953; Ahmadjian 1967; Nimis and Poelt 1987; Alstrup and Hawksworth 1990; Hawksworth 2003; Santesson et al. 2004; Lawrey and Diederich 2003). Besides these well documented lichenicolous fungi which are quite apparent on the upper cortex of lichen thallus, an abundant population of non-obligate microfungi was also revealed from within the lichens (Suryanarayanan et al. 2005). Majority of the accessory fungi reported from inside the lichen thalli i.e. endolichenic fungi (Miadlikowska et al. 2004a, b) are phylogenetically distinct from lichenicolous fungi and more closely related to member of ascomycetes found as endophytes in vascular plants. Most of the endolichenic fungi are restricted within the thallus (Miadlikowska et al. 2004a, b) and adapted to precise positions inside the thallus while some are restricted to particular strata, such as occurring in the medulla but not the algal layer (Sun et al. 2002). On the basis of the intra-ecology of the thallus, lichens are incredibly intricate, i.e. a body made as a result of symbiotic association of two primary algal and fungal partners and the participation of free living bacteria and non symbiotic fungi in this association all of which led us to consider lichens as a miniature ecosystem (Farrar 1976; Seaward 1988; Bjelland and Ekman 2005). Recently the presence of yeasts in lichens have opened a new horizon and made it a more complex system (Spribille et al. 2016).

Lichens are basically made up of an algal and a fungal partner, but the characteristic features of algal partner are not considered when a lichen species is described and the features of fungal partners are considered as the features of lichen species itself. When it comes down to classifying the lichens into natural systems its biggest strength "the symbiosis" becomes its worst enemy, since, they do not have their own phylogeny (Tehler and Irestedt 2007). To overcome this problem they are integrated into fungal system and are classified with chitinous fungi, and it is obvious that neither the life form nor the growth form are unique features to the symbiotic associations but have been subject for repeated gains and losses during the evolutionary history (Tehler and Irestedt 2007). Nearly all the lichenized fungi belong to ascomycota or rarely to basidiomycota, consequently known as ascolichens and basidiolichens. Further, the ascolichens mainly belong to Sordariomycetes, Lecanoromycetes and Eurotiomycetes. The Lecanoromycetes is almost entirely lichenized and also holds the tremendous population of lichen-forming species. Natural relationships within the Euascomycotina are beginning to be resolved, and major and minor groups are emerging, even though much work remains to be done (Nash 2008).

Lichens are among the most humble and least demanding poikilohydric organisms. Since, they have no mechanism to prevent desiccation, they desiccate and remain dormant when their environment dries out but can rehydrate when water becomes available again. They only need light, minute amounts of mineral nutrition from dust or rain and occasional humidity. They usually absorb water directly through their body surface by aerosol, mist and water vapors and due to this nature they live long in dry areas. Accordingly they are able to grow even on bare rocks, on bark of trees or suspended in the forest canopy, where water is only provided a short time daily or only during a short period of the year. Many species will survive extremely low temperatures and very harsh environmental conditions (Søchting 1999).

Lichens are perennial and slow growing organisms; the slow growth rate is mainly attributed to the growth of mycobiont. In lichens there is no central nutrient supply to the growing regions; on the contrary they have the onsite food production by the photobiont which is further used by both the algal and fungal partners. In general, lichens have centrifugal, apical and marginal growth pattern. Varied growth forms can be observed in different growth forms of lichens for example crustose and foliose lichens show radial growth on the other hand fruticose lichens grow in length. The annual growth rate in crustose lichens ranges from 0.2 to 1.0 mm, in foliose lichens it is 1.0 to 2.5 mm and 2.0 to 6.0 mm in fruticose lichens (Awasthi 2000).

Lichens display an extraordinary multiplicity of colors comprising black, brown, gray, green, orange, red and yellow (Wirth 1995; Brodo et al. 2001, Nash 2008) and size ranging from less than a mm^2 to long, pendulous forms that hang over 2 m from tree branches. Besides this, they also exhibit intriguing morphological variation in growth forms and are initially characterized into leprose, crustose, squamulose, foliose and fruticose. If the thallus is crustose, then their morphology may vary from species to species and can be categorized variously viz. continuous, rimose, areolate, rimose-areolate, bullate, verrucose, granular, indeterminate, determinate, effuse, irregular, orbicular, lobate, zonate, radiate, dendritic, pulvinate, floccose, byssoid, effigurate, placodioid (Nash et al. 2002). In case of foliose thallus, whether it is monophyllous or polyphyllous, variation in lobe shape viz. linear, regular, sublinear or subirregular, ligulate, lingulate, spathulate, flabellate, truncate, rotund; lobe margin shape viz. entire, irregular, flexuose, crenate, incised, dissected, laciniate, lacerate, sinuous, recurved, reflexed, crisped, revolute, involute, cucullate, corniculate, inflexed, deflexed; surface details viz. postulate, reticulate, tomentose, presence or absence of vegetative propagules, maculae, pseudocyphellae, veins, cilia, cephalodia, etc.; organ of attachment viz. rhizines – simple, squarrose, dichotomously branched, tufted; holdfast, hypothallus etc. are important taxonomic characters need to be searched for (Nash et al. 2002). In case of fruticose thallus whether it is circular or angular in cross section; pendulous, erect, decumbent or subfruticose; radiate, stratose radiate; branching pattern – isotomic/anisotomic/sympodial, appressed, decumbent, semi-erect, erect, pendent; branches – subulate, terete, torulose, tortuous, moniliform, ventricose, angulate, campanulate, alate; surfaces – foveolate, striation, presence of pseudocyphellae, vegetative propagules, papillose, fibrillose etc. are important taxonomic characters (Nash et al. 2002).

Among various growth forms of lichens in the evolutionary succession, leprose are considered pioneers, followed by crustose, placodioid, squamulose, foliose, dimorphic and fruticose being the latest. Leprose, crustose and some placodioid and squamulose lichens are called microlichens as they are smaller in size and mostly require a compound microscope for their identification. On the other hand, foliose, fruticose and dimorphic lichens are called macrolichens, as they have a comparatively larger thallus and can be identified with the help of hand lens and stereozoom microscope.

Like other group of plants of simpler organization, lichens also show vegetative (*fragmentation, lateral spinules, blastidia, phyllidia, schizidia, goniocysts, hormocysts, soredia, isidia*), asexual (*conidia*) and sexual (*ascospores* produced within apothecia and perithecia, and *basidiospores*) methods of reproduction. The so called sexual reproduction in lichens refers to the sexual reproduction of the mycobiont only.

Lichens are distributed among all the imaginable habitats worldwide, and can be found growing on almost all sorts of substrata, including both the natural as well as artificial ones. They are mainly found living epiphytically on tree barks (*corticolous*), on dead wood (*lignicolous*) and on twigs (*ramicolous*), and various species show a clear case of specificity towards the selection of trees to grow epiphytically on the basis of the physico-chemical properties of the bark of the plant. Some lichens prefer growing on "acid-rich" bark (e.g. spruce, birch or alder) while others on "base-rich" bark (e.g. walnut, Norway maple). Besides this, lichens often used to colonize mosses (*muscicolous*) and bare soil (*terricolous*); and act as a significant constituent of cryptogamic soil crusts in arid and semi-arid regions (Evans and Johansen 1999; Belnap and Lange 2003). Furthermore, lichens occur almost ubiquitously on rocks (*saxicolous*), either growing over the surface (*epiliths*) or embedded few millimeters deep inside the rock (*endoliths*) (Friedmann 1982). Selection of rock depends on the type of rock; some lichens prefer calcareous rocks (limestone or dolomite) while others lime-free silicate rocks (granite, gneiss, or basalt). Some fast growing lichens found in the tropics and subtropics are found colonizing the surface of leaves as epiphylls (*foliicolous*) (Lücking and Bernecker-Lücking 2002). Majority of lichens are terrestrial but there are some exceptions to this, *Peltigera hydrothyria* colonizes freshwater streams and some species of *Lichina* and *Verrucaria maura* group used to colonize marine intertidal zone. Prior to this, the man made substrates which lichens used to colonize includes rubber, plastic, glass, iron, clothes, stonework, concrete, plaster, ceramic, tiles, brick. Whatever be the substrate, they are somewhat restricted to particular substrate attributes, such as, "preferences" by definite light and moisture conditions, pH, texture and chemical composition. Their distribution on the same susbtrate is influenced by different environmental variables too (such as light, exposure, altitude, moisture and orientation) at multiple scales (Sheard and Jonesen 1974; Eversman 1982; Lesica et al. 1991; McCune and Geiser 1997; Crites and Dale-Mark 1998; Uliczka and Angelstam 1999; Brodo et al. 2001; Lehmkuhl 2004; Lalley et al. 2006). Many are "sun species", others "shade species"; many are restricted to cool, humid areas and are more often found in forests, others tolerate even sunny, dry conditions and live on freestanding trees and rocky outcrops. Some are able to live on rain-protected flanks and in bark crevices, while others are dependent upon the frequent soaking of the thallus with water (Wirth 1995).

As mentioned earlier that some lichens used to colonize only upon calcareous and base rich substrata, while others only on very acidic substratum, indicating that lichen colonization on a particular substrate is pH dependent, and following pH conditions were described by Wirth (1972):

S. No.	Substrate	pH value	Nature of substrate
1	*Extremely acidic*	3.3	Extreme acidophytic
2	*Very acidic*	3.4–4.0	Very acidophytic
3	*Rather acidic*	4.1–4.8	Rather acidophytic
4	*Moderately acidic*	4.9–5.6	Moderately acidophytic
5	*Sub neutral*	5.7–7.0	Sub neutrophic
6	*Neutral*	7.0	Neutrophic
7	*Moderately basic*	7.1–8.5	Moderately basic
8	*Basic*	over 7	Basic
9	*Embracing*	a wide pH range	Euryion

Later Gombert et al. (2004) also categorized lichen species into three groups as per type of bark and nutrient need – *acidophytic* (species which prefer to grow on acidic barks), *neutrophytic* (species which are not concerned about the bark type and nutrients) and *nitrophytic* (species which prefer deciduous tree bark enriched with dust or nutrients), – and this classification is generally employed as an ecological tool and standard feature to assess epiphytic lichen flora.

Not only substratum, but also vertical distribution and the large scale climatic conditions play an important role in the distribution of the species and one must always be conscious about these ecological data. Henceforth, following altitude readings, regarding distribution of lichens in a region, should be kept in mind while collecting lichens from a particular geographic region: (i) *Coastal* (including both littoral and sublittoral species), (ii) *Tropical*, (iii) *Subtropical*, (iv) *Foothills*, (v) *Submontane*, (vi) *Montane*, (vii) *Subalpine*, (viii) *Alpine*.

Lichens are not able to continue constant metabolic activity, for example, they cannot regulate their water budget. Because of the lack of true roots lichens are unable to take up water actively and since they have no evaporation protection system they slow down the water loss during draught conditions. During extreme draught situations lichens turn into an almost 'lifeless' organism as they allow to lose that much amount of water which is necessary and sufficient to sustain metabolism. Lichens are capable of taking up water through their upper surface like a sponge in a very short period of time. Whenever the lichen thallus gets some moisture it starts its metabolic process and continues as before draught. So this ability of lichens, pause and restart the metabolic process like a song in mp3 player, provides them the potential to survive in harsh conditions and it is also responsible for the extra-ordinarily slow development of lichens (Wirth 1995).

Many lichens produce secondary metabolites, so called lichen substances or lichen acids, as a byproduct of primary metabolism. These secondary metabolites which are of fungal origin comprising up to 20% of the dry weight of thallus (5–10% is common) are synthesized via three metabolic pathways: (1) acetyl-poly-malonyl pathway, (2) mevalonic acid pathway and (3) shikimic acid pathway (Boustie and Grube 2005) and are deposited extra-cellularly rather than inside the hyphae; that's why also called as extracellular products. The functions of secondary metabolites play a vital role in their distribution inside a lichen thallus e.g. majority of these metabolites are found in medulla and some are in cortex region. On the basis of the chemical structures, most secondary metabolites are depsides (barbatic acid), depsidones (salazinic acid), depsones (picrolichenic acid), dibenzofurans (usnic acid), lactones (protolichesterinic and nephrosterinic acid), phenolic compounds (orcinol and β-orcinol derivatives), pulvinic acid derivatives (vulpinic acid) and quinones (parietin) (Shukla et al. 2010). Some of these metabolites have pigments which can be generally found in the cortex region of the lichens thallus and are responsible for assigning wide range of colors e.g. brown, red, yellow or yellow-greenish in various lichens. However, majority of the secondary metabolites are colorless and almost all of them are localized in the medullary region. A large number of lichens produce some specific metabolites (may be out of the reason that they got them genetically or acquired them during the process of adaptation to live in certain specific places in an ecosystem) which come in handy in the taxonomy (Lawrey 1977) and referred as chemo-taxonomy (Hawksworth 1976; Frisvad et al. 2008). Therefore, the knowledge of secondary metabolites of lichens is very significant as sometimes it helps in the determination of a species. The lichen substances can be identified through High Performance Liquid Chromatography (HPLC), microcrystallography, 'spot test' and Thin Layer Chromatography (TLC). The classical spot color test technique (K, C, Pd, I) helps to get preliminary idea about the possible extrolite present which involve application of specific reagents directly on the lichen thallus. Thin Layer Chromatography (TLC) is also extensively used with standardized protocols affording more accurate information on metabolic profiles, while microcrystallography involves identification of lichen compounds from the shape of crystals observed under microscope. Modern spectrometric techniques involving hyphenated techniques (GC-MS, LC-MS/MS etc.), coupling separation and identification of lichen compounds enable high throughput characterization of bioactive compounds. Besides this, many substances – corresponding as well to the lichen parts containing these substances – glow white, blue-white, yellow, or other colors, in the light of commercial UV-lamps, hence, UV-lamps having short wave (e.g. 254 nm) and long wave (e.g. 366 nm) can be used for demarcating species. All these tests are routinely or more frequently used for identification of secondary metabolites and hence lichen taxa.

Since lichens lack cuticle or epidermis and are devoid of a well-developed root system, therefore they not only absorb nutrients directly from the atmosphere, but also absorb and/or adsorb pollutants, and didn't show any visible signs of injury. Lichens show differential sensitivity towards wide range of pollutants; some species are inherently more sensitive, while others show tolerance to high levels of pollutants. In general, air pollution sensitivity increases among growth forms in the following series: crustose < foliose < fruticose, though there are some exceptions to this gradation (Shukla et al. 2014). Considerable longevity, wide distribution, slow growth, perennial nature, uniform morphology, surface structure and roughness, are some salient features of lichens which makes them the best and sensitive bioindicators or early warming indicators of forest health and ecological continuity as well as atmospheric pollution in various regions across the globe (McCune 2000; Kricke and Loppi 2002; Brunialti and Giordani 2003; Wolseley et al. 2006; Shukla et al. 2014). Hawksworth (1971) rightly said that lichens are undoubtedly reliable bioindicators in monitoring ecosystem changes and act as Litmus test for ecosystem health. Biomonitoring provides relevant information about ecosystem health either from changes in the behavior of the monitored organism (species composition and/or richness, physiological and/or ecological performance, morphology) or from the concentrations of specific substances in the monitoring organisms (Shukla et al. 2014). The most commonly used lichen biomonitoring methods are *community analysis, lichen tissue analysis, lichen zone mapping, sampling individual species* and *transplant studies*. Christopher et al. (2015) in their lichen community approach (Box 2) observed that when phanerogams are scattered throughout large areas of forest and large sample area becomes essential to express the diversity of the forest, then in that case lichen communities in a small area can reflect the range of habitats available, species richness, and amount of disturbance or degradation of the forest, indicating that these forest lichen communities respond to primary climate variables such as precipitation and temperature and to geographical gradients such as elevation and latitude that integrate climate factors (Will-Wolf et al. 2002, 2006; Nash 2008). To scale up species composition to more generalized ecological theories rather than focusing on species specific conclusions, Nelson et al. (2015) used Trait-based approach in determining lichen community composition, as Trait patterns along environmental gradients have been interpreted as indicators of mechanisms behind lichen adaptations to the environment (Stanton and Horn 2013; Färber et al. 2014; Nelson et al. 2015). Hence, due to the quality of being the best bio-indicator of air-quality and forest health, lichens are being employed globally to identify significant biodiversity-rich sites which need an appropriate management approach of forest resources and to designate the regions which require conservation (Will-Wolf 2010). Besides, lichen diversity can be utilized to assess the forest health and status in regions where other forms of environmental monitoring are expensive or impractical (Box 2; Wolseley and Aguirre-Hudson 2007). The identification of 'indicator' lichens (Table 1) can provide a basis for management recommendations and can also be used to assess climatic change and potential forest recovery in areas where deforestation has caused a change in local climate and phanerogamic communities (Wolseley and Aguirre-Hudson 2007).

Garty et al. (2003) classified lichens into three categories based on their responses to air pollution – *sensitive species*, with varying degrees of sensitivity to the detrimental effects of pollutants, but ultimately succumbing to air pollution; *tolerant species*, resistant to pollution, belonging to the native community and remaining intact in their native habitat; and *replacement species*, making their

Box 2 Different Epiphytic Lichen Communities and Their Indicator Values

Group		Community	Indicators	Remarks
I	Early successional (Pioneer) communities of smooth-barked mesotrophic microhabitats	**Type A:** *Arthonia radiata, Lecidella elaeochroma* Community	***Arthonia radiata,*** *Buellia disciformis, Lecanora chlarotera,* ***Lecidella elaeochroma,*** *Pertusaria leioplaca*	Associated with younger and smooth-barked broad leaf trees with higher bark pH Type B Community appears highly restricted and sensitive to macroclimatic settings, such that it is marginally less common in the climatically outlying driest sites (where it intergrades with Type A) or wettest sampled habitats (where it intergrades with Type G). In contrast, the Type A *Arthonia radiata-Lecidella elaeochroma* Community is broadly distributed, though with a microhabitat preference for moderately shaded and drier local environments, e.g. away from the higher humidity associated with water courses.
		Type B: *Graphis scripta* Community	*Arthonia didyma,* ***Graphis scripta,*** *Pertusaria hymenea, Pyrenula occidentalis*	
II	Early successional to mature communities in mesotrophic microhabitats	**Type C:** *Frullania dilatata* Community	***Frullania dilatata,*** *Ulota bruchii, Ulota crispa*	Occur on younger broadleaf trees, though persisting (for the *Frullania dilatata* Community), and increasing (for the *Phlyctis argena-Ramalina farinacea* Community) on older and mature trees. The Type D *Phlyctis argena-Ramalina farinacea* Community is a dominant community type in mesotrophic microhabitats in relatively more continental climates, while the Type C *Frullania dilatata* Community is geographically more widespread and occurs more frequently than Type D in oceanic systems.
		Type D: *Phlyctis argena- Ramalina farinacea* Community	*Melanelixia glabratula* agg., *Parmelia sulcata, Lepra amara (=Pertusaria amara),* ***Phlyctis argena,*** *Orthotrichum affine,* ***Ramalina farinacea***	

(continued)

Box 2 (continued)

Group		Community	Indicators	Remarks
III	Mature communities in drier microhabitats (rough-barked and/or leaning trees)	**Type E:** *Calicium viride-Chrysothrix candelaris* Community	*Anisomeridium biforme, Arthonia vinosa,* ***Calicium viride,*** *Chaenotheca trichialis,* ***Chrysothrix candelaris,*** *Cliostomum griffithii, Lepraria incana*	These represent drier microhabitats, both within deep furrows and on sheltered undersides of large leaning broadleaf trees. The Type E *Calicium viride-Chrysothrix candelaris* Community is more sensitive to macroclimate and is common in relatively continental settings. In contrast, the Type F *Lecanactis abietina* Community is widespread geographically, but is sensitive to additional microhabitat factors including canopy openness and distance to water, and locally favors drier and more open stand conditions.
		Type F: *Lecanactis abietina* Community	***Lecanactis abietina***	
IV	Mature mesotrophic communities in oceanic climates (or humid microclimates)	**Type G:** *Lobaria virens-Normandina pulchella-Metzgeria furcata* Community	*Frullania fragilifolia, Isothecium alopecuroides, Lepraria eburnea,* ***Lobaria virens, Metzgeria furcata, Normandina pulchella,*** *Opegrapha vulgata, Thelotrema lepadinum, Zygodon viridissimus*	It is the oceanic climate equivalent to Community Type D and is more frequent in relatively warmer microhabitats such as sheltered south-facing wooded slopes. It occurs in mesotrophic broadleaf habitats on relatively young trees. In ecological terms it is a later-successional element that develops from the early successional Type B and Type C Community, within the oceanic setting.

(continued)

Box 2 (continued)

Group		Community	Indicators	Remarks
V	Late successional mesotrophic communities in oceanic climates (or humid microclimates)	**Type H:** *Hypnum cupressiforme* agg.- *Usnea flammea* Community	***Hypnum cupressiforme*** agg., *Lepraria lobificans, Plagiochila punctata,* ***Usnea flammea***	This group is dominated by bryophytes as well as larger fruticose and foliose lichens, and characterizes later successional epiphyte communities in milder and/or oceanic climates. The Group V alliance may occur in leached or slightly more oligotrophic habitats than the more mesotrophic Group IV Community Type G. The Group V Community types show additional sensitivity to a range of microhabitat factors, including the angle of bole lean for Type H and Type I Community, as well as local humidity, expressed as distance to water, for Community Type K.
		Type I: *Hypnum andoi-Microlejeunea ulicina* Community	*Cladonia coniocraea, Dicranum scoparium,* ***Hypnum andoi,*** *Lepraria rigidula,* ***Microlejeunea ulicina***	
		Type J: *Frullania tamarisci* Community	***Frullania tamarisci,*** *Harpalejeunea molleri*	
		Type K: *Lobaria pulmonaria-Isothecium myosuroides* Community	*Hypotrachyna taylorensis,* ***Isothecium myosuroides, Lobaria pulmonaria,*** *Parmotrema crinitum, Parmotrema perlatum*	
VI	Early successional to mature communities in intermediate settings	**Type L:** *Arthopyrenia cinereopruinosa-Lecanora pulicaris* Community	***Arthopyrenia cinereopruinosa,*** *Chrysothrix flavovirens,* ***Lecanora pulicaris,*** *Micarea micrococca* agg., *Pertusaria pupillaris*	It is a very loosely associated group of crustose lichens, each of which may occur interspersed within a mosaic of more mature and competitive (foliose/bryophyte dominated) epiphyte communities. It is commonest within oligotrophic woodland settings that characterize sites in the cool and wet central Highland belt, located between a warmer and more humid oceanic zone, and a cooler and drier north-eastern continental zone. However, it is not strongly restricted in terms of macroclimate, and can be geographically widespread. Locally, it occurs in drier microclimatic settings as opposed to the constant humidity of watercourses.

(continued)

Box 2 (continued)

Group		Community	Indicators	Remarks
VII	Mature to late successional communities in oligotrophic microhabitats	**Type M:** *Hypotrachyna laevigata-Loxospora elatina* Community	*Anisomeridium ranunculosporum,* ***Hypotrachyna laevigata, Loxospora elatina,*** *Mycoblastus caesius, Scapania gracilis*	Characterizes relatively more acid-barked and oligotrophic microhabitats. It includes the transition between Community Type M in more oceanic climatic settings and potentially occurring on younger trees, and the later successional Community Type N on older trees. Community Type M is most likely to occur within mesotrophic microhabitats. Type N also occupies an intermediate climatic position most strongly associated geographically with the cool and wet central Highland belt. This is in contrast to the more continental Community Type O.
		Type N: *Mycoblastus sanguinarius-Protoparmelia ochrococca-Sphaeorophorus globosus* Community	*Cladonia macilenta/ polydactyla, Micarea synotheoides,* ***Mycoblastus sanguinarius,*** *Ochrolechia androgyna, Parmelia saxatilis* agg., *Platismatia glauca,* ***Protoparmelia ochrococca, Sphaerophorus globosus,*** *Usnea subfloridana*	
		Type O: *Bryoria fuscescens-Ochrolechia microstictoides-Parmeliopsis hyperopta* Community	***Bryoria fuscescens,*** *Xylospora friesii (=Hypocenomyce friesii), Hypocenomyce scalaris, Hypogymnia physodes, Imshaugia aleurites, Lecidea hypopta, Lecidea nylanderi, Lepraria jackii* agg., ***Ochrolechia microstictoides, Parmeliopsis hyperopta,*** *Pertusaria borealis, Nephromopsis chlorophylla (= Tuckermanopsis chlorophylla), Violella fucata (Mycoblastus fucatus)*	

Source: Christopher et al. (2015)

Table 1 Various lichen communities indicating different set of environmental conditions

S. No.	Lichen communities	Indicators
1	Calicioid community	Undisturbed old forest ecosystem.
2	Alectoroid and Usnioid community	The tufted and pendulous fruticose lichens including genera *Sulcaria, Bryoria, Ramalina* and *Usnea* comes under this community, and are indicators of old forest with better air quality.
3	Cyanophycean community	The variation in diversity and abundance of epiphytic cyanolichens appears useful as an indicator of forest ecosystem function and used to indicate forest age and continuity (McCune 1993) and play an important role in forest nutrient cycle.
4	Lobarian community	The Lobarian group comprised of *Lobaria, Pseudocyphellaria, Peltigera* and *Sticta* indicates species rich old forest with long forest continuity (Gauslaa 1995) and are very sensitive to air quality.
5	Xanthoparmelioid community	Indicates stable productive landscape i.e. landscapes with no accelerated erosion or least trampling by animals and trekking by humans.
6	Graphidioid and Pyrenuloid community	It consists of members of Graphidaceous (*Graphis, Opergrapha, Sarcographa, Phaeographis*) and Pyrenocarpous (*Anthracothecium, Pyrenula, Lithothelium, Porina*) lichens and indicates young and regenerated forest since they prefer to grow on a smooth bark tree in evergreen forest/regenerated forest.
7	Lecanorioid community	This group comprised of *Lecanora, Lecidella* and *Biatora* prefer to grow on trees in thinned-out, regenerated or disturbed forests with more open area to receive more light and wind and indicates well-illuminated environmental condition of the forest with considerable exposure of light and wind.
8	Parmelioid community	Members of this group comprised mostly the species of lichen genera *Bulbothrix, Flavoparmelia, Parmotrema, Parmelia, Punctelia* and other genera of Parmeliaceae. On one hand some of its members indicate closed canopy forests receiving less sunlight, while on the other hand some species indicates open thinned-out forest with more sunlight. In nut shell they indicate thinned out forest.
9	Pertusorioid community	The group includes species of lichen genus *Pertusaria* and indicates old tree forest with rough-barked trees.
10	Leprarioid community	The species of *Chrysothrix, Cryptothecia* and *Lepararia* are the common lichens of the Leprarioid group, which forms powdery thallus on the substrates. They used to indicate moist and dry vertical slopes, rough barked trees of moist and dry habitats. Besides this, the species of *Chrysothrix* appears first after forest fire.
11	Lecideoid community	The members of the group such as *Lecidea, Protoblastedia, Haematomma, Bacidia, Buellia* and *Schadonia* used to colonize mostly bark of deciduous trees in sheltered and well-lit exposed sides and hence indicate exposed illuminated area.

(continued)

Table 1 (continued)

S. No.	Lichen communities	Indicators
12	Physcioid community	The lichen species of *Physcia, Pyxine, Dirinaria, Heterodermia, Phaeophyscia* and *Rinodina* belongs to this group and indicates polluted and nitrophilous environment as these lichens are considered as pollution tolerant lichens.
13	Teloschistacean community	The species of *Caloplaca, Letroutia, Brigantiaea* and *Xanthoria* having yellow thallus and apothecia belongs to this group and indicates high UV irradiance since the members of this group have an ability to grow both on exposed and sheltered rocks and the dark orange pigment present on the upper cortex of the thallus acts as a filter and protects the lichens from high UV radiation.
14	Dimorphic community	Species of the genera *Cladonia, Cladina* and *Stereocaulon* forms this community and indicates undisturbed soil ecosystem.
15	Lichinioid community	The genera of the lichen family Lichiniaceae mostly having cyanobacteria as their photobiont belongs to this group and indicates presence of calcareous substrates in the habitats as the members of this group prefer to colonize dry rocks and barks having higher concentration of calcium.
16	Peltuloid community	The species of lichen genera *Peltula* belongs to this group and indicates a stable rock substratum.

Source: Shukla et al. (2014)

appearance after destruction of the major part of the native lichen community as a result of pollution.

Though they are very slow growing and their annual harvesting is not economically feasible, yet are economically important and have been used as food (e.g. *Cetraria islandica, Umbilicaria*), fodder (e.g. *Cladonia rangiferina, Ramalina fraxinea*), medicines (e.g. *Lobaria pulmonaria, Xanthoria parietina, Peltigera canina, Usnea*), dyes (e.g. *Roccella montagnei, Ochrolechia tartarea, Parmelia saxatilis, Letharia vulpina, Teloschistes flavicans*) and perfumery (e.g. *Evernia mesomorpha, Evernia prunastri, Pseudevernia furfuracea*). The biological activities viz. antioxidant, antimicrobial and cytotoxicity, antiviral, antiretroviral, anti-inflammatory, analgesic, etc. carried out by lichens is due to lichen substances. Lichen secondary metabolites also act as allelochemicals, which protect lichens from herbivores grazing. The cortical accumulation of lichen compounds increases opacity of the upper cortex, restraining solar irradiance reaching the light sensitive algal layer. Various light screening pigments display strong UV absorption abilities and might function as filters in order to prevent excessive UV-B irradiation. Some lichen phenolics have high ability to scavenge toxic free radicals generated by UV light. Some other metabolites step in metal homeostasis and pollution tolerance (Shukla 2017). Not only they are economically important, but are ecologically important too (Shukla 2017). They are the pioneer colonizers of rocks and boulders which are exposed by receding glaciers, landslides or by volcanic eruptions, and hence cause weathering of rock surface. Not only they used to weather the rock surface, but their presence in rocks and boulders which are exposed by receding glaciers can be used for knowing glacier retreat by performing lichenometric studies. Lichens have several miscellaneous uses too, such as, poisoning the wolves (e.g. *Letharia vulpina*, *Vulpicida pinastri*), tanning of leather (e.g. *Cetraria islandica*, *Lobaria pulmonaria*), stuff material for dolls and pillows (e.g. species of *Bryoria*, *Usnea*).

Although studies carried on lichens to characterize their potential bioactive constituents is increasing, but it is restricted by difficulties encountered in identification of the species, collection of bulk quantities, and the isolation of pure substances, which limit the number of tested bioactive compounds. Moreover, detection and isolation of minor compounds is generally restricted by the high abundance of redundant lichen compounds, which may overcome involving highly sophisticated high throughput techniques which allow characterization of more potential interesting bioactive compounds. In contrast to lichens, endolichenic and lichenicolous fungi till date are untapped resource of potential biomolecules. To some extent it is believed that endolichenic fungi contribute significantly to the synthesis of lichen metabolites, hence, can be used as an interesting organism for biological studies (Nayaka 2017), and by this we can to some extent conserve the lichen diversity from being harnessed.

Since lichens play an important role in forest ecosystems, such as, contributing to forest biodiversity (Dettki and Esseen 1998; Kuusinen and Siitonen 1998; Lesica et al. 1991; Pharo et al. 1999), used as forage by many animals (Rosentreter et al. 1997; Zabel and Waters 1997), provide nesting material for birds (Hayward and Rosentreter 1994; Starkey and Hagar 1999), constitute preferred habitat for many invertebrates (Pettersson et al. 1995), involved in nutrient cycling (Boucher and Nash 1990; Esseen et al. 1996; Knops et al. 1991; Pike 1978), but their number is declining day by day and many of the lichen species are becoming rare, as their living area due to the activities of human being is either being destroyed or all too frequently is subjected to change. The evidence for this decline is produced by old publications about the lichen flora, from others still maintaining access to collections of lichens from an earlier time. In old herbaria species are represented, which no longer exist today; many lichens extremely rare today are present from places, where they in the meantime have certainly disappeared. The reasons for the decline of lichens are many folds, such as, (1) decline in the percentage of protected areas, (2) rapid loss of forest area, either due to deforestation or conversion of forest area into cropland, (3) forest fire, (4) monoculture plantation practices, (5) secondary forest plantations, (6) cutting down of trees along avenues and streets, (7) habitat alteration by doing afforestation with exotic tree species, (8) air pollution by power stations, industry, domestic combustion and traffic. Therefore, environmental changes result in alteration of habitats and ecosystems at local, regional as well as global scale resulting in loss of lichen biodiversity; extinction of sensitive species and invasion of thermophilic species.

These creatures are very important in our lives, but we know very little about them. In comparison to other plants and wild animals, they have received poor media coverage and are being overlooked for their role as important bioindicators. The impact of this knowledge is profound, and has led to lichens being largely ignored in conservation management and policy throughout the world. Likewise fungi, there is also critical shortage of experts in their classification and identification. Besides this lack of accessible information on rarely recorded species may merit conservation. Except in a very few cases, we do not know whether these species are genuinely rare or merely overlooked during the survey work. Conservation activities are highly dependent on baseline data, backed up by analysis of communities and survey and monitoring of species at risk. In case of lichens, identification manuals do exist, but cover only a proportion of the better known species and in general focus on field characters which are frequently inadequate for accurate recording without more technical analysis. So what could be the possible ways for conserving lichens? The strategies suggested by Moore et al. (2001) for fungal conservation could be applied in case of lichens too, which includes (1) protection of the environment and thereby the natural community itself, i.e. conservation of habitats, (2) conservation of mycologists/lichenologists, as they can make serious contributions to knowledge of species sufficiently needed to be quickly conserved, (3) change in general perceptions of the society towards microbes as they rank very low in kinship scale i.e. increase in kinship with the microbes (Staley 1997) which can be achieved by public education. In other words lichenologists should make use of the popularity of lichens among people, (4) in-situ conservation of non-mycological or non-lichenological reserves/ecological niches, (5) fungus/lichen favorable land management practices, (6) inclusion of mapping programs and ecological data during survey, (7) legal protection of lichens, either in national or international legislation, (8) funding of research and international cooperation in favor of lichen conservation.

References

Acharius E (1798) Lichenographiae Svecicae Prodromus. D.G. Björn, Linköping. https://doi.org/10.5962/bhl.title.79420

Acharius E (1803) Methodus qua omnus detectos Lichenes. F.D.D. Ulrich, Stockholm. https://doi.org/10.5962/bhl.title.79411

Acharius E (1810) Lichenographia universalis. Dandewerts, Göttingen. https://doi.org/10.5962/bhl.title.79418

Acharius E (1814) Synopsis methodica lichenum. Lund

Ahmadjian V (1967) The lichen symbiosis. Blaisdell Publishing, Toronto/Waltham, MA, pp 250

Ahmadjian V (1993) The lichen symbiosis. Wiley, New York

Ahmadjian V, Jacobs JB (1983) Algal-fungal relationships in lichens: recognition, synthesis, and development. In: Goff LJ (ed) Algal symbiosis. Cambridge University Press, New York, pp 147–172

Alstrup V, Hawksworth DL (1990) The lichenicolous fungi of Greenland. Meddelelser om Grönland, Bioscience 31:1–90

Armaleo D, Clerc P (1990) Lichen chimeras: DNA analysis suggests that one fungus forms two morphotypes. Exp Mycol 15:1–10

Armstrong R (2004) Lichens, lichenometry and global warming. Microbiol:32–35

Aschenbrenner IA, Cernava T, Berg G, Grube M (2016) Understanding microbial multi-species symbioses. Front Microbiol 7(1–9):180

Awasthi DD (2000) A hand book of lichens. Bishan Singh Mahendra Pal Singh, Dehradun

Barreno E, Herrera-Campos M, García-Breijo F et al (2008) Non photosynthetic bacteria associated to cortical structures on Ramalina and Usnea thalli from Mexico. [WWW document]. URL http://192.104.39.110/archive/IAL6abstracts.pdf. Asilomar, Pacific Grove, CA, USA: Abstracts IAL 6-ABLS Joint Meeting, pp 5
Beck A (2002) Photobionts: diversity and selectivity in lichen symbioses. Int Lichenol Newslett 35:18–24
Beck A, Friedl T, Rambold G (1998) Selectivity of photobiont choice in a defined lichen community: inferences from cultural and molecular studies. New Phytol 139:709–720
Belnap J, Lange OL (2003) Biological soil crusts: structure, function and management. In: Baldwin IT, Caldwell MM, Heldmaier G et al (eds) Ecological studies series. Springer, Berlin, pp 1–503
Bjelland T, Ekman S (2005) Fungal diversity in rock beneath a crustose lichen as revealed by molecular markers. Microb Ecol 49:598–603
Boonpragob K, Crittenden PD, Lumbsch TH (2012) Lichens: from genome to ecosystems in a changing world. MycoKeys 6:1–2
Boucher VL, Nash TH III (1990) The role of the fruticose lichen *Ramalina menziesii* in annual turnover of biomass and macronutrients in a blue oak woodland. Bot Gaz 151(1):114–118
Boustie J, Grube M (2005) Lichens-a promising source of bioactive secondary metabolites. Plant Genetic Resources 3(2):273–287
Brodo IM, Sharnoff SD, Sharnoff S (2001) Lichens of North America. Yale University Press, New Haven
Brunialti G, Giordani P (2003) Variability of lichen diversity in a climatically heterogeneous area (Liguria, NW Italy). Lichenologist 35:55–69
Cambridge dictionary. https://dictionary.cambridge.org/us/dictionary/english/lichen
Christopher JE, Sally E, Marios T et al (2015) Epiphyte communities and indicator species. An ecological guide for Scotland's woodlands. Royal Botanic Garden, Edinburgh
Casano LM, del Campo EM, García-Breijo FJ et al (2011) Two *Trebouxia* algae with different physiological performances are ever-present in lichen thalli of Ramalina farinacea. Coexistence versus competition? Environ Microbiol 13(3):806–818
Cobanoglu G, Sesal C, Gokmen B et al (2010) Evaluation of the antimicrobial properties of some lichens. South-west J Hortic Biol Environ 1(2):153–158
Collins dictionary. https://www.collinsdictionary.com/dictionary/english/lichen
Crites S, Dale-Mark RT (1998) Diversity and abundance of bryophytes, lichens and fungi in relation to wood substrate and successional stage in aspen mixed wood boreal forests. Can J Bot 76:641–651
De Bary A (1879) Die Erscheinung der Symbiose. Verlag Karl Trübner, Strasbourg
de Jussieu A (1730) De la nécessité d'établir dans la méthode nouvelle des plantes, une classe particulière pour les fungus. Hist Acad Roy Sci Mém Math Phys 1730:377–383
des Abbayes H (1953) Travaux sur les lichens parus de 1939 à 1952. Bulletin Sociétié Botanique de France 100:83–123
Dettki H, Esseen PA (1998) Epiphytic macrolichens in managed and natural forest landscapes: a comparison at two spatial scales. Ecographysiol 21:613–624
Elenkin A (1902) Zur frage der Theorie des Endosaprophytismus bei Flechten. Bull Jard Imp St Petersb 2:65–84
Esseen PA, Renhorn KE, Pettersson RB (1996) Epiphytic lichen biomass in managed and old-growth boreal forests: effect of branch quality. Ecol Appl 6(1):228–238
Evans RD, Johansen JR (1999) Microbiotic crusts and ecosystem processes. CRC CR Rev Plant Sci 18(2):183–225
Eversman S (1982) Epiphytic lichens of a ponderosa pine forest in southeastern Montana. The Bryologist 85:204–213
Färber L, Solhaug KA, Esseen P et al (2014) Sunscreening fungal pigments influence the vertical gradient of pendulous lichens in boreal forest canopies. Ecology 95(6):1464–1471
Farrar JF (1976) The lichen as an ecosystem: observation and experiment. In: Brown DH, Hawksworth DL, Bailey RH (eds) Lichenology: progress and problems. Academic Press, London, pp 385–406

Forman RTT (1975) Canopy lichens with blue-green algae: a nitrogen source in a Columbian rain forest. Ecology 56(5):1176–1184

Friedl T (1989) Systematik und Biologie von Trebouxia (Microthamniales, Chlorophyta) als Phycobiont der Parmeliaceae (lichenisierte Ascomyceten). Ph.D. thesis. Universtät Bayreuth, Bayreuth

Friedmann EI (1982) Endolithic microorganisms in the Antarctic cold desert. Science 215:1045–1053

Frisvad JC, Anderson B, Thrane U (2008) The use of secondary metabolite profiling in chemotaxonomy of filamentous fungi. Mycol Res 112:231–240

Galloway DJ (1992) Biodiversity: a lichenological perspective. Biodivers Conserv 1:312–323

Galun M, Kardish N (1995) Lectins as determinants of symbiotic specificity in lichens. Cryptog Bot 5:144–148

Garty J, Tomer S, Levin T et al (2003) Lichens as biomonitors around a coal-fired power station in Israel. Environ Res 91:186–198

Gauslaa Y (1995) The Lobarion, an epiphytic community of ancient forests, threatened by acid rain. Lichenologist 27:59–76

Gerson U, Seaward MRD (1977) In: MRD S (ed) Lichen-invertebrate associations. Lichen ecology. Academic Press, London, pp 69–120

Gombert S, Asta J, Seaward MRD (2004) Assessment of lichen diversity by index of atmospheric purity (IAP), index of human impact (IHI) and other environmental factors in an urban area (Grenoble, southeast France). Sci Total Environ 324:83–199

Grube M, Cardinale M, de Castro JV et al (2009) Species-specific structural and functional diversity of bacterial communities in lichen symbioses. ISME J 3(9):1105–1115

Hale ME, Cole M (1988) Lichens of California. University of California Press, Berkeley, CA

Hawksworth DL (1971) Lichens as litmus for air pollution: a historical review. Int J Environ Stud 1:281–296

Hawksworth DL (1976) Lichen chemotaxonomy. In: Brown DH, Hawksworth DL, Bailey RH (eds) Lichenology: progress and problems. Academic, London, pp 139–184

Hawksworth DL (1988) The variety of fungal-algal symbioses, their evolutionary significance, and the nature of lichens. Bot J Linn Soc 96:3–20

Hawksworth DL (2003) The lichenicolous fungi of Great Britain and Ireland. An overview and annotated checklist. Lichenologist 35:191–192

Hawksworth DL, Honegger R (1994) The lichen thallus: a symbiotic phenotype of nutritionally specialized fungi and its response to gall producers. In: Williams MAJ (ed) Plant galls: organisms, interactions, populations. Clarendon Press, Oxford, pp 77–98

Hayward GD, Rosentreter R (1994) Lichens as nesting material for northern flying squirrels in the northern Rocky Mountains. J Mammal 75:663–673

Henssen A, Jahns HM (1973 [1974]) Lichenes. Eine Einführung in die Flechtenkunde. Stuttgart: Thieme

Honegger R (1988) The functional morphology of cell-to-cell interactions in lichens. In: Scannerini S, Smith DC, Bonfante-Fasolo P et al (eds) Cell to cell signals in plant, animal and microbial symbiosis, Springer/NATO ASI series Berlin, Series H, vol 17, pp 39–53

Honegger R (1991) Functional aspects of the lichen symbiosis. Ann Rev Plant Phys Plant Mol Biol 42(1):553–578

Honegger R (2000) Great discoveries in bryology and lichenology – Simon Schwendener (1829–1919) and the dual hypothesis of lichens. Bryologist 103:307–313

Ihda TA, Nakano T, Yoshimura I et al (1993) Phycobionts isolated from Japanese species of *Anzia* (lichenes). Arch Protistenkd 143:163–172

Jovan S (2008) Lichen Bioindication of biodiversity, air quality, and climate: baseline results from monitoring in Washington, Oregon, and California. General technical report, PNW-GTR-737. Portland. U.S. Department of Agriculture, Forest Service, Pacific Northwest Research Station. pp 115

Kirk PM, Cannon PF, Minter DW et al (2008) Dictionary of the Fungi, 10th edn. CABI, Wallingford

Knops JMH, Nash TH III, Boucher VL et al (1991) Mineral cycling and epiphytic lichens: implications at the ecosystem level. Lichenologist 23:309–321
Kricke R, Loppi S (2002) Bioindication: the IAP approach. In: Nimis PL, Scheidegger C, Wolseley PA (eds) Monitoring with lichens – monitoring lichens. Kluwer, Dordrecht, pp 21–37
Kuusinen M, Siitonen J (1998) Epiphytic lichen diversity in old-growth and managed *Picea abies* stands in southern Finland. J Veg Sci 9:283–292
Lalley JS, Viles HA, Copeman N et al (2006) The influence of multi-scale environmental variables on the distribution of terricolous lichens in a fog desert. J Veg Sci 17:831–838
Lawrey JD (1977) Adaptive significance of *O*-methylated lichen depsides and depsidones. Lichenologist 9:137–142
Lawrey J, Diederich P (2003) Lichenicolous fungi: interactions, evolution, and biodiversity. The Bryologist 106:80–120
Lehmkuhl JF (2004) Epiphytic lichen diversity and biomass in low-elevation forests of the eastern Washington Cascade range, USA. Forest Ecol Manag 187:381–392
Lesica P, McCune B, Cooper SV et al (1991) Differences in lichen and bryophyte communities between old-growth and managed second growth forests in Swan Valley, Montana. Can J Bot 69:1745–1755
Linnaeus C (1753) Species plantarum. Salvius, Stockholm
Lücking R, Bernecker-Lücking A (2002) Distance, dynamics, and diversity in tropical rainforests: an experimental approach using foliicolous lichens on artificial leaves. I. Growth performance and succession. Ecotropica 8:1–13
McCune B (1993) Gradients in epiphyte biomass in three *Pseudotsuga-Tsuga* forests of different ages in western Oregon and Washington. The Bryologist 96:405–411
McCune B (2000) Lichen communities as indicators of forest health. The Bryologist 103:353–356
McCune B, Geiser L (1997) Macrolichens of the Pacific Northwest. Oregon State University Press, Corvallis
Medlin JJ (1996) Michigan Lichens. In: Cranbrook Institute of Science. Bloomfield Hills, Michigan
Merriam Webster dictionary. https://www.merriam-webster.com/dictionary/lichen
Miadlikowska J, Arnold AE, Hofsteler V et al (2004a) High diversity of cryptic fungi inhabiting healthy lichen thalli in a temperate and tropical forest, In: Randlanc T, Saag A (eds) Lichens in Focus. Book of Abstracts of the 5th IAL Symposium, Tartu University Press, Estonia, pp 43
Miadlikowska J, Arnold AE, Lutzoni F (2004b) Diversity of cryptic fungi inhabiting healthy lichen thalli in a temperate and tropical forest. Ecol Soc Am Annu Meet 89:349–350
Mitchell ME (2007) Signposts to symbiosis. A review of early attempts to establish the constitution of lichens. Huntia 13(2):101–120
Moore D, Nauta MM, Evans SE et al (2001) Fungal conservation – issues and solutions. Cambridge University Press, New York, pp 262
Muggia L, Grube M (2018) Fungal diversity in lichens: from extremotolerance to interactions with algae. Life 8(12). https://doi.org/10.3390/life8020015
Nash TH III (2008) Lichen biology, II edn. Arizona State University, Cambridge University Press, New York, pp 484
Nash III TH, Ryan BD, Gries C et al (2002) Lichen flora of the Greater Sonoran desert region, Vol 1, Lichens Unlimited, Arizona State University, Tempe, AZ, pp 532
Nayaka, S (2017) Future prospects in lichenology in Indian contest. In: Sinha GP (ed) Workshop manual for Workshop on Lichen Identification and Nomenclature, BSI, Allahabad, Rajesh Corporation, pp 9–11
Nelson PR, McCune B, Swanson DK (2015) Lichen traits and species as indicators of vegetation and environment. The Bryologist 118(3):252–263
Nieboer E, Richardson DHS, Tomassini FD (1978) Mineral uptake and release by lichens: an overview. The Bryologist 81(2):226–246
Nienburg W (1917) Über die Beziehungen zwischen den Algen und Hyphen im Flechtenthallus. Zeitschrift für Botanik 9:529–545
Nimis PL, Poelt J (1987) The lichens and lichenicolous fungi of Sardinia (Italy). An annotated list. Studia Geobotanica Trieste 7(suppl.1):1–269

Oxford Living Dictionaries. https://en.oxforddictionaries.com/definition/lichen
Pettersson RB, Ball JP, Renhorn KA et al (1995) Invertebrate communities in boreal forest canopies as influenced by forestry and lichens with implications for passerine birds. Biol Conserv 74:57–63
Pharo EJ, Beattie AJ, Binns D (1999) Vascular plant diversity as a surrogate for bryophyte and lichen diversity. Conserv Biol 13:282–292
Pike LH (1978) The importance of epiphytic lichens in mineral cycling. The Bryologist 81:247–257
Poelt J (1973) Systematic evaluation of morphological characters. In: Ahmadjian V, Hale ME (eds) The Lichens. Academic Press, London
Rambold G, Triebel D (1992) The inter-lecanoralean associations. Biblioth Lichenol 48:3–201
Rambold G, Friedl T, Beck A (1998) Photobionts in lichens: possible indicators of phylogenetic relationships? The Bryologist 101:392–397
Reinke J (1872) Ueber die anatomischen Verhältnisse einiger Arten von *Gunnera* L. *Nachrichten von der Königl.* Gesellschaft der Wissenschaften und der Georg-Augusts-Universität zu Göttingen 9:100–108
Rosentreter R, Hayward GD, Howard MW (1997) Northern flying squirrel seasonal food habits in the interior conifer forests of central Idaho, USA. Northwest Sci 71:97–102
Sanders WB (2001) Lichens: the Interface between mycology and plant morphology. Bioscience 51(12):1025–1035
Santesson R, Moberg R, Nordin A et al (2004) Lichen–forming and lichenicolous fungi of Fennoscandia. Museum of Evolution, Uppsala University, Uppsala
Scharnagl K (2019) The scale of symbiosis. Symbiosis. https://doi.org/10.1007/s13199-019-00601-x
Schwendener S (1867) Über die wahre Natur der Flechtengonidien. Verhandlungen der schweizerischen naturforschenden Gesellschaft 57:9–11
Seaward MRD (1988) Contribution of lichens to ecosystems. In: Galun M (ed) CRC handbook of lichenology. CRC Press, Boca Raton, pp 107–129
Sheard JW, Jonesen MW (1974) A multivariate analysis of the distribution of lichens on *Populus lemuloids* in West-Central Canada. The Bryologist 77:514–530
Shukla V (2017) Lichen secondary metabolites and their uses. In: Sinha GP (ed) Workshop manual for workshop on lichen identification and nomenclature, BSI, Allahabad, Rajesh Corporation, pp 19–21
Shukla V, Upreti DK, Patel DK et al (2010) Accumulation of polycyclic aromatic hydrocarbons in some lichens of Garhwal Himalayas, India. Int J Environ Waste Manag 5(1/2):104–113
Shukla V, Upreti DK, Bajpai R et al (2014) Spatial distribution of metallic content in Himalayan ecosystem monitored using lichens. Water Air Soil Pollut 224:1859
Søchting U (1999) Lichens of Bhutan – biodiversity and use. Botanical Institute, University of Copenhagen, Denmark, pp 29
Speer BR, Ben W (1997) Lichens: life history & ecology. University of California Museum of Paleontology Last modified 2006. http://www.ucmp.berkeley.edu/fungi/lichens/lichenlh.html
Spribille T, Tuovinen V, Resl P et al (2016) Basidiomycete yeasts in the cortex of ascomycete macrolichens. Science 353:488–492 https://doi.org/10.1126/science.aaf8287
Staley JT (1997) Biodiversity: are microbial species threatened? Curr Opin Biotech 8:340–345
Stanton DE, Horn HS (2013) Epiphytes as 'filter-drinkers': Life-form changes across a fog gradient. The Bryologist 116:34–42
Starkey E, Hagar JC (1999) Biodiversity of young forest-role of birds. Final Report to US Department of Interior, Blue River District
Sun HJ, Depriest PT, Gargas A et al (2002) *Pestalotiopsis maculans*: a dominant parasymbiont in North American lichens. Symbiosis 33:215–226
Suryanarayanan TS, Thirunavukkarasu N, Hariharan GN et al (2005) Occurrence of non-obligate microfungi inside lichen thalli. Sydowia 57(1):20–130
Tehler A, Irestedt M (2007) Parallel evolution of lichen growth forms in the family Roccellaceae (Arthoniales, Euascomycetes). Cladistics 23:432–434
The Free dictionary. https://www.thefreedictionary.com/lichen

Tournefort JP (1694) Elements de Botanique. Imprimerie Royale, Paris

Uliczka H, Angelstam P (1999) Occurrence of epiphytic macrolichens in relation to tree species and tree age in managed boreal forest. Ecography 22(4):395–404

Voss EG, Burdet HM, Chaloner WG et al (1983) International code of botanical nomenclature, adopted by the thirteenth international botanical congress, Sydney, August 1981. Regnum Veg 111: i–xv, 1–472

Will-Wolf S (2010) Analyzing lichen indicator data in the forest inventory and analysis program. General technical report, PNW-GTR-818. U.S. Department of Agriculture, Forest Service, Pacific Northwest Research Station, Portland, pp 62

Will-Wolf S, Neitlich P, Esseen PA (2002) Monitoring biodiversity and ecosystem function: forests. In: Nimis PL, Scheidegger C, Wolseley P (eds) Monitoring with lichens–monitoring lichens. NATO Science Series. Kluwer Academic Publishers, The Hague, pp 203–222

Will-Wolf S, Geiser LH, Neitlich P et al (2006) Comparison of lichen community composition with environmental variables at regional and subregional geographic scales. J Veg Sci 17:171–184

Wirth V (1972) Die Silikatflechten – Gemeinschaften im ausseralpinen Zentraleuropa. Dissertationes Botanicae 17:1–306

Wirth V (1995) Die Flechten Baden-Württembergs, 2nd edn Ulmer, Stuttgart

Wolseley PA, Aguirre-Hudson B (2007) Lichens as indicators of environmental changes in the tropical forests of Thailand [Online] http://www.jstor.org/locate/envpol.

Wolseley PA, Stofer S, Stofer R et al (2006) Variation of lichen communities with land use in Aberdeenshire, UK. Lichenologist 38:307–322

Zabel CJ, Waters JR (1997) Food preference of captive northern flying squirrels from the Lassen National Forest in northeastern California. Northwest Sci 71:103–107

Zahlbruckner A (1907) Lichenes (Flechten). In: Engler A, Prantl K (eds) Die Naturlichen Pflanzenfamilien. W. Engelmann, Leipzig, pp 49–249

Zahlbruckner A (1926) Afrikanische Flechten (Lichenes). Engl Bot Jahrb 60:468–552

Zopf W (1897) Über Nebensymbiose (Parasymbiose). Ber deut Bot Ges 15:90–92

Introduction to Endophytic Fungi Associated with Lichens i.e. Endolichenic Fungi

Abstract With an estimated diversity of 5.1 million species, fungal kingdom is the most diverse group of eukaryotes on the blue planet but only a handful have been described leaving majority of the diversity missing. The missing fungal species can be found in three categories which are (1) fungi in tropical forests; (2) **fungi in unexplored habitats**; (3) lost or hidden species. The endophytes fall into the second category and considered to play a crucial role in the progression from an aquatic to a terrestrial way of life. Endophytes can be defined as "***fungi that live within their host plants without causing any noticeable symptoms of disease and may become pathogenic when the host got stressed***". They are ubiquitous, highly diverse and are estimated at least 400 million years old. Their life-history strategy varies from facultative saprobe, to parasitic, to exploitive to mutualistic. Because of their specific functions and survival strategy in the host tissues, they produces secondary metabolites having novel biochemistry which belong to various chemical groups i.e., aliphatics, alkaloids, cytochalasines, depsipeptides, furandiones, isocumarines, phenols, quinines, steroids, terpenoids and xanthones. About 80% of these compounds are biologically active and have been commercially consumed for pharmaceutical and agricultural purposes. Endophytes have been isolated from the plants growing in boreal, temperate, tropical, xeric environments, extreme arctic, alpine regions and from almost all plant lineages including lichens. Generally lichens are considered as a classic symbiotic involvement of a fungus and an alga but it is found not to be completely true as lichens also harbor some other asymptomatic fungi called endolichenic fungi (ELF). Endolichenic fungi are species-rich among the Pezizomycotina (Ascomycota) and are predominant from the classes such as Dothideomycetes, Leotiomycetes, Pezizomycetes and Sordariomycetes. Till date over 35 endolichenic microorganisms have been cultivated and studied in detail which leads to the chemical characterization of 196 novel natural product structures out of a total of 351 secondary metabolites, showing a wide range of biological activities. In comparison to the endophytes, the endolichenic fungi (their diversity, ecology, biology, physiology, nutrition, relationship with the host etc.) are still in the dark. This chapter arises many significant logical questions regarding endolichenic fungi which suggest the need to study this less studied group of fungi.

M. Tripathi, Y. Joshi, *Endolichenic Fungi: Present and Future Trends*,
https://doi.org/10.1007/978-981-13-7268-1_2

The fungal kingdom is one of the most diverse groups of eukaryotes on earth (Blackwell 2011; Hibbett and Taylor 2013) and the whole fungal wealth is projected around 5.1 million (Blackwell 2011) of which only about 100,000 have been documented (Blackwell 2011; Kirk et al. 2008). Hawksworth and Rossman (1997) recognized following three places where we can found those missing fungi: (1) fungi in tropical forests; (2) **fungi in unexplored habitats**; (3) lost or hidden species.

The endophytes which used to fall in the second category i.e. fungi in unexplored habitats are thought to be one of the most essential evolutionary steps from an aquatic to a terrestrial lifestyle (Selosse and Le Tacon 1998; Heckman et al. 2001). They are ubiquitous and highly diverse, with the reported majority being members of Pezizomycotina (Ascomycetes) (Arnold et al. 2009), of which, many lack a known teleomorphic state (Carroll 1988). Most studies with the help of culture-based methods have highlighted that the high proportion of endophytes belong to Dothideomycetes, Leotiomycetes, Pezizomycetes, and Sordariomycetes (Petrini and Petrini 1985; Petrini et al. 1990; Bacon and White 2000; Stone et al. 2004; Arnold and Lutzoni 2007; Shipunov et al. 2008; Arnold et al. 2009). But at this juncture, one should keep in mind that types of media used in the study also have an immense effect on the representation of fungal classes, e.g. use of malt extract agar as the isolation medium gives a low population of Eurotiomycetes (Arnold et al. 2009; U'Ren 2011; U'Ren et al. 2010, 2012). This result also corroborates numerous culture-dependent (Lodge et al. 1996; Higgins et al. 2007; Shipunov et al. 2008) and culture-independent investigations (Zimmerman and Vitousek 2012; U'Ren et al. 2014).

The term "Endophyte" (from Greek "endo" – within and "phytos" – plant) was coined by de Bary (1866) including each and every individual from virulent foliar pathogens to mycorrhizal root symbionts. Later other workers (Carroll 1986; Petrini 1991; Wilson 1995; Bacon and White 2000; Schulz and Boyle 2005) redefined the terminology and excluded mycorrhizal and other fungi occurring both outside as well as inside a plant from the endophyte category (Box). According to the most prevalent definition, fungal endophytes are ***fungi that live within their host plants without causing any noticeable symptoms of disease*** (Carroll 1988; Petrini 1991; Wilson 1995; Stone et al. 2000) ***and may become pathogenic when the host got stressed*** (Carroll 1988). Some even speak of the '*true endophytes*', meaning those whose colonization never results in visible disease symptoms (Mostert et al. 2000).

The symbiotic interactions between plants and fungi are estimated around 400 million years old (Remy et al. 1994; Krings et al. 2007). Fossil evidences of plant-associated microbes suggest that microbes are living as endophytes from the first appearance of higher plants on Earth (Zhang et al. 2006). As mentioned earlier, de Bary (1866) introduced the term endophyte but it was Guerin (1898) who for the first time reported that these creatures are harmless inhabitants. The very first attempt of describing complete life of *Lolium temulentum* colonized with the fungal endophyte was made by Freeman (1904). Among all microbial endophytes the fungal endophytes grabbed more attention rather than bacterial or other microbial endophytes since they are more applicable. The year 1977 turned out to be the landmark year in the history of endophytic research because of the discovery of endophytic fungus *Epichloe coenophiala* (=*Neotyphodium coenophialum*) from *Festuca arundinacea* which is the cause of "fescue toxicosis" (Bacon et al. 1977). Later on, it was established that toxicity in the plant was caused by the toxic alkaloid "peramin" produced by the fungus *N. coenophialum*. Another important discovery in the field of endophytic research was the discovery of an anticancer drug "taxol", from *Taxomyces andreanae*, an endophytic fungus isolated from *Taxus brevifolia* (Stierle et al. 1993). This led to the search for endophytes worldwide from each and every group of organisms for a better understanding of their ecological function and capability to create prospective bioactive natural products for the betterment of the human society.

The life-history strategy of endophytes is tremendously discussed by various workers and found to be very inconsistent, varying from facultative saprobe, to parasitic, to exploitive, to mutualist (Saikkonen et al. 1998; Schulz and Boyle 2005). On the other hand Rosenblueth and Martínez-Romero (2006) gave a defined representation to the endophytic way of life by categorizing them into two subgroups, viz. (a) *obligate:* those which have complete dependency on the metabolism of host for their survival and are being spread amongst hosts by the activity of different types of vectors or by vertical transmission (Hardoim et al. 2008) and, (b) *facultative:* those which live outside the host body during a certain stage of their life cycle and are mostly associated with plants from its neighboring soil environment and atmosphere (Abreu-Tarazi et al. 2010). Endophytes have been implicated in decreased herbivory, increased tolerance against drought, salinity, heat, metals etc., and enhancement of plant growth (Fröhlich et al. 2000; Schardl et al. 2004; Sieber 2007; Redman et al. 2011). In woody perennials they are thought to protect the plants in which they live by one or more mechanisms (antibiosis, mycoparasitism, induced resistance and/or competitive exclusion), and are thought to develop from environmental or background inoculum and are not transferred from generation to generation (Johnson and Whitney 1992). Therefore, plants that have been removed from their natural environment and cultivated are thought to become depleted in their specific or coevolved endophytes (Taylor et al. 1999) and, as a result, may become more susceptible to pests and diseases.

During last two decades, it has been observed that a great deal of microbial diversity having novel biochemistry and secondary metabolite production dwell in plant tissues (Strobel 2006; Porras-Alfaro and Bayman 2011). Endophytes are metabolically proactive than their free counterparts since they have specific functions in nature and have various active metabolic pathways to survive in the host tissues (Strobel and Daisy 2003; Strobel 2006; Riyaz Ul-Hassan et al. 2012). A prominently elevated proportion of fungal endophytes (80%) manufacture secondary metabolites possessing biologically active compounds (Schulz et al. 2002) synthesized via various metabolic pathways (Tan and Zou 2001). These secondary metabolites belonging to diverse structural groups i.e., aliphatics, alkaloids, cytochalasines, depsipeptides, furandiones, isocumarines, phenols, quinines, steroids, terpenoids and xanthones, have been commercially utilized for pharmaceutical, medical and agricultural purposes (Tan and Zou 2001; Castillo et al. 2002, 2003; Strobel and Daisy 2003; Ezra et al. 2004; Li et al. 2005; Park et al. 2005; Gunatilaka 2006; Wang et al. 2007; Suryanarayanan and Shaanker 2009; You et al. 2009). Some of these metabolites represent novel structural groups, e.g. the palmarumycins (Kröhn et al. 1997) and a new benzopyroanone (Kröhn et al. 2002). The proportion of novel structures produced by endophytes (51%) is considerably higher than that produced by soil isolates (38%), demonstrating that endophytes are certainly a superior supply of novel secondary metabolites (Schulz et al. 2002). In some cases, the endophytic fungi are proficient to yield similar bioactive metabolites as the host plant itself, and the excellent illustration of this is the production of taxol by endophytic fungi *viz. Taxomyces andreanae* and *Pestalotiopsis microspora* isolated from *Taxus brevifolia* (Strobel et al. 1996; Li et al. 1998).

The endophytic colonization of land plants by fungi is ubiquitous and the endophyte–host relationship is hypothesized to be multifaceted (*loc. cit.*) and may show a discrepancy from plant to plant and microbe to microbe depending upon the state of affairs. Endophytic microbes enter in host tissue in same way as the pathogenic microorganisms do enter into the plants. But, because of their feeble pathogenic behavior, the plant produces defensive molecules in a less significant amount in the vicinity of infection which does not create obstacles in the entry of endophytes when compared to pathogens. Endophyte after infection primarily can inhabit asymptomatically within the host tissue, but later either may become latent pathogen and produce disease symptoms during adverse conditions or may become latent saprophytes, showing antagonistic behavior on the other end of the continuum.

Endophytes have been isolated from plants found in alpine, boreal, temperate and tropical forests, including extreme arctic (Petrini 1987; Fisher et al. 1995), and xeric environments (Mushin and Booth 1987; Mushin et al. 1989); and from mesic temperate and tropical forests. Endophytic fungi have been obtained from several plant lineages e.g. algae (Cubit 1974; Hawksworth 1988; Kröhn et al. 2005; Wang et al. 2006; Yang et al. 2006; Thirunavukkarasu et al. 2011; Mathan et al. 2013), bryophytes (Döbbler 1979; Pocock and Duckett 1985; Ligrone 1988; Ligrone et al. 1993; Chambers et al. 1999; Davis et al. 2003; Kauserud et al. 2008; U'Ren et al. 2010; Zhang et al. 2013), pteridophytes (Petrini et al. 1992; Fisher et al. 1992; Schmid and Oberwinkler 1993; Swatzell et al. 1996; Sati and Belwal 2005; Sati et al. 2009; Kumaresan et al. 2006, 2013), gymnosperms (Carroll and Carroll 1978; Carroll and Petrini 1983; Sahashi et al. 1999; Tan and Zou 2001; Strobel and Daisy 2003; Sieber 2007; Rodriguez et al. 2009; Thongsandee et al. 2012), angiosperms (Sydowia 1914; Clay 1991; Elmi and West 1995; Janardhanan and Ahmad 1997; Saikkonen et al. 1998; Rajagopal and Suryanarayanan 2000; Clay and Schardl 2002; Arnold et al. 2003; Malinowski et al. 2004; Nalini et al. 2005; Raviraja 2005; Tejesvi et al. 2005; Gond et al. 2007; Lin et al. 2007; Verma et al. 2007; Kharwar et al. 2008, 2010, 2011; Mishra et al. 2014; Verma et al. 2014) but their existence within lichens was reported recently (Petrini et al. 1990; Girlanda et al. 1997; Suryanarayanan et al. 2005, 2017; Li et al. 2007; Tripathi et al. 2014a, b, c, d; Tripathi and Joshi 2015; Wang et al. 2016; Maduranga et al. 2018). Based on phylogeny and life history traits, they have been classified into two broad groups: a) **Clavicipitaceous** - includes endophytes infecting some grasses confined to cool regions and, b) **Non-clavicipitaceous** - includes endophytes, which are from asymptomatic tissues of non-vascular plants, ferns and allies, conifers and angiosperms and are limited to the Ascomycota or Basidiomycota group (Jalgaonwala et al. 2011; Bhardwaj and Agrawal 2014).

Table 1 The range of fungal-algal symbioses involving different numbers of bionts (Hawksworth 1988)

Number of Bionts	Examples
Two-biont symbioses	
Mycobiont as inhabitant	Mycophycobiosis
	Fungal parasites of algae
Mycobiont as exhabitant	Lichens
Three-biont symbioses	
Two photobionts: One mycobiont	Cephalodia
	Blue-green/green morphotypes
	Algicolous lichens
	Bryophyllous lichens
Two mycobionts: One photobiont	Lichenicolous fungi
	Mechanical hybrids
Four-biont symbioses	
Three photobionts: One mycobiont	Cephalodia
Two photobionts: Two mycobionts	Lichenicolous lichens
Three mycobionts: One phycobiont	Fungi on lichenicolous fungi
Five- or more biont symbioses	Mechanical hybrids

For about over a century lichens were just considered a classic case of mutualism, but alternatively, lichens were also deemed as a case of controlled parasitism, since the fungus appears to acquire majority of the benefits and the photobiont may grow in a steady pace in the lichenized state than when free-living (Ahmadjian 1993). Now the recent researches are enlightening these modest organisms to be astonishingly complex.

Regardless of the established concept about lichens as bipartite- (mycobiont with algal or cyanobacterial photobiont) or tripartite symbioses (mycobiont with algal and cyanobacterial photobionts) (Table 1), lichen thalli additionally harbor other organisms and this is substantiated by Boonpragob et al. (2012) where they stated that lichens are not only regarded as an individual, but also as a '*functional organismic community*' or as a microhabitat with a vast array of coexisting fungal, algal and bacterial genotypes.

Since lichens are the pioneer colonizers of an ecosystem which means that they were the first living entity to be available as hosts for other microorganisms in comparison to any other plant, hence, consequently they could be the evolutionary source for the parasitic/pathogenic/saprotrophic fungi. They make available a true, although still mostly unexplored, ecological niche for a wide variety of microorganisms, including bacteria and non-culturable non-photosynthetic bacteria (Gerson and Seaward 1977; Cardinale et al. 2008; Hodkinson and Lutzoni 2009; Grube et al. 2015; Biosca et al. 2016; Muggia et al. 2016) and eukaryotes (e.g., fungi, arthropods, and nematodes) (Bates et al. 2011; Park et al. 2014). Among all these, fungi exhibit interactions with the lichen host ranging from parasites to commensals to diversely dependent saprobes (Hawksworth 1979, 1980, 1981, 1982, 1983;

Honegger 1996; Jeffries and Young 1994), and their distribution depends on their ecological and physiological features, such as resistance to environmental stress and tolerance to secondary metabolites produced by most of their hosts (Lawrey 1995; Torzilli and Lawrey 1995). Fungi other than the primary mycobiont, and possibly not mere passive contaminants, are frequently encountered during direct observation of lichen thalli, and their competitive presence often makes it difficult to isolate the mycobiont (Crittenden et al. 1995; Honegger 1996). Direct examination methods inescapably propose a constrained depiction of the natural status, while techniques of isolation may obstruct the peculiarity between fungi that are actual lichen-inhabitants and those present as resting stages. Successful culture of single lichen-associated fungi has sporadically been accounted (Gams 1971; Hawksworth 1975, 1979, 1981; Hawksworth and Jones 1981; Crittenden et al. 1995). Only a few studies, however, has been set out to explore the fungal communities that can be isolated from lichen thalli (Petrini et al. 1990; Girlanda et al. 1997; Li et al. 2007; Arnold et al. 2009; Kannangara et al. 2009; U'Ren et al. 2010).

The fungi besides the obligate fungal partner residing inside/outside the healthy lichen thalli are referred to as accessory or secondary fungi (considering the fungal partner as a primary fungus) and comprises incidental fungi on thallus surface, lichenicolous fungi, and endolichenic fungi.

Hofstetter et al. (2007) were the first persons to suggest that non-lichen fungal sequences commonly amplified from lichens may represent a previously unrecognized source of lichen-specific fungal diversity, which were later referred to as "endolichenic" forms by Miadlikowska et al. (2004) and Arnold et al. (2009). Arnold et al. (2009) viewed them as phylogenetically distinct from both lichen-forming mycobionts and lichenicolous fungi and an important evolutionary link to plant-associated endophytes (Fleischhacker et al. 2015; Chagnon et al. 2016; Muggia et al. 2016; U'Ren et al. 2010, 2012, 2014).

The endolichenic fungi have been discovered within living, apparently healthy lichen thalli by various workers (Petrini et al. 1990; Girlanda et al. 1997; Suryanarayanan et al. 2005; Li et al. 2007; Arnold et al. 2009) and are known to coexist with the mycobiont (which forms the bulk of the lichen thallus) and live in very close association with the photobiont (the algal or cyanobacterial partner in the lichen symbiosis) (Arnold et al. 2009). They are entirely different from the lichen mycobionts (the primary fungal component of the lichen thallus), lichenicolous fungi (which fruit or are otherwise symptomatic on thalli), and incidental fungi on thallus surfaces (Lutzoni et al. 2001; Lawrey and Diederich 2003; Arnold et al. 2009).

Endolichenic fungi are analogous to the plant endophytes in having some similarities in many aspects *viz.* (1) they inhabit the intercellular spaces of the hosts, (2) do not produce any visible disease symptoms, (3) are transmitted horizontally and (4) produce an array of secondary metabolites (Arnold et al. 2009; Kannangara et al. 2009; U'Ren et al. 2012). Likewise endophytes, endolichenic fungi are also species-rich among the Pezizomycotina (Ascomycota) (Arnold et al. 2009) and are predominant from the classes such as the Dothideomycetes, Leotiomycetes, Pezizomycetes and Sordariomycetes (Petrini and Petrini 1985; Petrini et al. 1990; Bacon and White 2000; Stone et al. 2004; Arnold and Lutzoni, 2007; Shipunov et al. 2008; Arnold et al. 2009). The members of endolichenic fungi are quite common among all major primary nonlichenized lineages of Euascomycetes (Dothideomycetes, Leotiomycetes, Pezizomycetes and Sordariomycetes) but are absent among the lichen-dominated clades (Lecanoromycetes, Arthoniomycetes, Lichinomycetes) and rare among the secondarily nonlichenized Eurotiomycetidae and Chaetothyriales.

Since very few studies are available on endolichenic fungi and hence it would be too early to say something about the association of endolichenic fungi with its host lichen. But to us it appears to be facultative in comparison to; and this can be justified by the fact that a lichen species, for example, *Parmotrema reticulatum*, when collected from five different localities was screened for the endolichenic fungal diversity (Tripathi and Joshi 2015) was found having different fungal species which did not match with each other. If the endolichenic fungi and lichens had had the obligate relationship then in each and every screening of same lichen species the exact fungal population should have appeared in the culture as in the previous one, but this was not the case. Thus, if they are not in an obligate symbiotic relationship with the host then whether they are in a mutualistic facultative relationship with the host needs to be answered in near future? It has been observed in case of endophytes of higher plants that the facultative mutualistic relationship depends on the host plants health condition, if the plant is healthy then the endophytes live in mutualistic partnership and when the plant faces stress (abiotic or biotic) the endophyte may change itself into a parasite (till the plant lives) or into a saprophyte (when the plant dies). This to some extent is also true in case of lichens where it has been found that some fungal species fall in the category of endolichenic (mutualistic) as well as lichenicolous fungi (parasitic) (for example, *Acremonium lichenicola* and *Fusarium* sp.). Fernández-Mendoza et al. (2017) in their study also revealed that lichens with and without obvious symptoms of infection harbor numerous described lichenicolous species in their mycobiomes. Hence, there may be a possibility that the present lichenicolous fungi are the endolichenic fungi of past on a weak lichen thallus which is facing some stress. Since very few described lichenicolous species have been sequenced till date, the ability to identify endolichenics as lichenicolous species is limited at present. However, as more sequences of described species are published, it will be possible to identify those previously assigned to endolichenic fungi that actually belong to described lichenicolous species (Diederich et al. 2018). However, the dearth of endolichenic fungi in Chaetothyriales corresponds with Arnold et al. (2009) in that lichenicolous fungi and endolichenic fungi are often phylogenetically dissimilar.

Besides this, studies carried out by Lawrey and Diederich (2003) support the idea that lichenicolous fungi are well-represented in both lichen-forming and non-lichenized clades of fungi. Since phylogenetic placement helps to elucidate origins of these fungi, it is apparent that lichenicolous fungi have multiple origins from lichen-forming and non-lichenized ancestors while endolichenic fungi are most likely descended from non-lichenized ancestors (Arnold et al. 2009). Endolichenics and lichenicolous represent components of the same communities and the distinction between them is likely to become less clear-cut as more lichen mycobiomes are studied and the inhabitants identified (Diederich et al. 2018).

Endolichenic fungi stand for an evolutionary incubator for conversions to endophytic associations in plants. Endolichenic-to-endophytic transitions occurred twice as frequently as transitions from endophytism to endolichenism (Arnold et al. 2009). Additionally, it is in concurrence with the results that endolichenic fungi are unusual colonizers of higher plants (Suryanarayanan et al. 2005). By keeping in view the huge diversity of lichenicolous fungi and the phylogenetic relationship between endolichenic fungi and endophytic fungi, it can be stated that lichens should also be made a vital part of fungal diversity estimation studies.

Because of the lack of reproductive bodies on the host plant the generation time of endophytic and endolichenic fungi is directly proportional to the life-span of the host plant. Persistent leaves and lichens would restrict the fungi living as endophytic and endolichenic symbionts to longer generation times in comparison to saprophytic fungi, possibly providing a strong selective pressure against transitions or reversions to symbiosis. Transitions from endolichenism to endophytism would be favored given the shorter generation times of leaves (< 1–15 years) relative to long-lived lichen thalli (up to >100 years). Endophytism could remain a viable strategy if endophytes act as ecological opportunists that form pathogenic (actively reproductive) infections in susceptible hosts while persisting as nonvirulent symbionts in other hosts (Arnold et al. 2009).

Lichens are very slow growing organisms having a lot of economic and ethnomedicinal value. They are being screened for their bioactive potential since time immemorial as they possess more than 1000 secondary metabolites. But because of their slow-growing nature, they cannot be harnessed on the industrial level to make drugs out of them. As mentioned earlier that endophytes produce similar metabolites like that of the host plant (Strobel et al. 1996), and if this holds true in case of lichens then there will be no need to harvest the lichen diversity, one just has to grow and ferment some endolichenic fungi.

Endolichenic fungi live a cryptic lifestyle inside lichen thalli and necessitate rather particular techniques for highest recovery (Petrini et al. 1990; Girlanda et al. 1997). Petrini et al. (1990) isolated 506 fungal taxa from 17 fruticose lichens, the majority of which (306) were isolated only once (Petrini et al. 1990). A further detailed study of lichen species *Xanthoparmelia taractica* (=*Parmelia taractica*) and *Peltigera praetextata* discovered contrasts in their fungal assemblages but similar levels of biodiversity (Girlanda et al. 1997). The highly porous and heterogeneous nature of the lichen thalli might be the reason behind the enormous fungal diversity it contains.

Investigations recovering endolichenic fungi are over a decade old practice (Petrini et al. 1990; Girlanda et al. 1997), but studies focused on isolating secondary metabolites from endolichenic fungi are more contemporary. The foremost research to isolate an endolichenic fungus was conducted by Petrini et al. (1990), who isolated filamentous fungi from sterilized segments of fruticose lichens *Cladonia* and *Stereocaulon*. They isolated a total of 506 fungal strain types; 166 of which were isolated more than once.

Girlanda et al. (1997) also used two foliose lichens (*Parmelia taractica* and *Peltigera praetexta*) to study the variety of fungal assemblages present in these lichens and obtained a total of 117 fungal isolates. Their study corroborates with

Petrini et al. (1990) in having most of the members belonging to the Pezizomycotina, Ascomycota.

Li et al. (2007) examined endolichenic fungi from members of five lichen families from the Baihua mountain of Beijing, China and reported 32 taxa from 488 segments of lichen thalli. The endolichenic fungi isolated by them belong to the phylum Ascomycota.

Arnold et al. (2009) in a recent investigation on phylogenetic estimation about the trophic transitions of Ascomycetes, presented some appealing approach onto the origin and evolution of endophytism and endolichenic fungi. Their study showed that endolichenic fungi or endolichenism have a significant role in the evolution of endophytism within the most species rich-phylum, Ascomycota. The findings of their investigation revealed that endolichenic fungi represent a swift evolutionary pace for fungal transitions to endophytic associations in plants. Their study was also the first to document that endolichenic fungi live inside the lichen thalli in very close association with the photobiont (Arnold et al. 2009).

U'Ren et al. (2010) investigated communities of endophytic fungi in mosses and endolichenic fungi in lichens using ITS rDNA sequences. They screened ten lichen species for endolichenic fungi and found that lichen species belonging to order Lecanorales, Peltigerales and Teloschistales showed the greatest isolation frequencies.

Cheon et al. (2013) not only isolated endolichenic fungi from four lichens *viz.* *Stereocaulon* sp. 1 (1429 strains), *Stereocaulon* sp. 2 (1430 strains), *Pezicula* (=*Cryptosporiopsis*) sp. (0156 strains) and *Graphis* sp. (1245 strains) but also examined 571 endolichenic fungal strains for their antifungal properties. Similarly, some other studies (Kannangara et al. 2009; Hwang et al. 2011; Kim et al. 2012; Padhi and Tayung 2015) have tried to uncover the endolichenic fungi with their compelling bioactivity, but since they worked on the crude extracts of endolichenic fungi, hence, complete identification of the metabolites is not yet available.

Recently Chen et al. (2015) described the novel order of fungi, Phaeomoniellales within the Eurotiomycetes and placed some endolichenic fungi within it.

U'Ren et al. (2016) utilized multigene phylogenetic analysis and reported that several isolates of endolichenic fungi and fungal endophytes obtained from the continental United States might represent novel species within the Xylariaceae, which need further study. The authors also concluded that both symbiotrophic and saprotrophic fungi reside within the Xylariaceae, which is one of the largest and most diverse families within the Pezizomycotina, Ascomycota (U'Ren et al. 2016).

Maduranga et al. (2018) isolated a total of 171 endolichenic fungal strains from lichens collected from mangrove and mangrove associated plants in Puttalam lagoon, Sri Lanka. They also investigated the effectiveness of ethyl acetate extracts of the endolichenic fungi isolates against antioxidant activity, antilipase activity and α-amylase inhibition activity in in-vitro conditions, and their results revealed that the extracts of *Daldinia eschscholtzii*, *Diaporthe musigena* and *Sordaria* sp. had the highest radical scavenging activity with smaller IC50 values (25–31 μg/mL) compared to the IC50 values of Butylated hydroxyl toluene (76.50 ± 1.47 μg/mL). The antilipase assay revealed that 13 extracts from endolichenic fungi showed promising antiobesity activity ranging between 25% and 40%. Amylase inhibitory assay carried out by them indicated that the test extracts do not contain antidiabetic secondary metabolites.

Till date over 35 endolichenic microorganisms have been cultured and subjected to detailed investigations leading to the chemical characterization of 196 novel natural product structures out of a total of 351 secondary metabolites, many of which have been shown to have a variety of biological activities (Table 5.1).

Suryanarayanan et al. (2005) for the first time studied five corticolous macrolichens (four foliose and one fruticose) for non-obligate microfungi residing inside the lichen thalli in India and used four different surface sterilization procedures. In addition to isolating endolichenic fungi, Suryanarayanan et al. (2005) also wanted to know whether endolichenic fungi were analogous to those occurring as endophytes within the bark and leaves (phellophytes) of the trees from which the lichens were collected. After the publication of Suryanarayanan et al. (2005), there exists a gap of 9 years and publications dealing with endolichenic fungi started appearing after 2014. Tripathi et al. (2014a, b, c, d) and Tripathi and Joshi (2015) isolated 25 endolichenic fungal isolates from 14 macrolichens of Kumaun Himalaya and most of the isolates belonged to Ascomycota and Zygomycota (basal fungal lineages). Vinayaka et al. (2016) and Jayakumar et al. (2016) isolated endolichenic fungi from some lichen taxa, but there is a little bit of hesitation in considering their results. Vinayaka et al. (2016) mentioned that they isolated 30 endolichenic fungi from 11 lichen taxa which they collected in the year 2013–2014. Though they mentioned that endophytes from lichens were isolated within 24 h of collection, but the online publication of the article in 2016 (which was also communicated in 2016) led us in accepting their data with some hesitation. Similarly, Jayakumar et al. (2016) isolated a single species of endolichenic fungi from one crustose lichen species which according to them was *Lecanora* sp. Their study was dubious in the sense that the host lichen was wrongly identified and also the isolation of a single species as endophyte raises the question that *whether single lichen thallus can host only one endophyte*? and according to previous studies, this could not happen. The number of endolichenic fungi isolated from a thallus can be countless and depend upon the number of replicates and the protocol which a worker follows. Shanmugam et al. (2018) not only isolated 10 endolichenic fungal strains from 06 different lichen species but also evaluated the biological potential of secondary metabolites isolated from these fungal strains which yielded promising results.

Besides these sporadic publications on Indian endolichenic fungi, so far only few review articles are available on this emerging branch mentioning the importance of this group of fungi (Singh et al. 2017; Suryanarayanan and Thirunavukkarasu 2017).

Initial investigations revealed that endolichenic fungi may be similarly species-rich to the endophytic fungi with which they co-exist in the few sites explored till date (Arnold et al. 2009; U'Ren et al. 2010) and that they are equally potential for pharmaceutical, agricultural, and industrial uses (e.g. Paranagama et al. 2007; Ding et al. 2009). In spite of increased curiosity about their diversity and applications, however, a great deal of things remains in the dark concerning the scale of endophytic and endolichenic biodiversity, and how communities are structured at the local level across assemblages of phylogenetically diverse hosts in natural ecosystems.

1 Conclusion

Fungal endophytes are now well recognized microbial flora and have been isolated from every category of plants (algae to angiosperms) for their prospective utilization in the benefit of humankind. In view of the fact that, a large gap is present between estimated fungal diversity (5.1 million) and described number of fungi (99,000), studying endophytic fungal diversity lightens a new ray of hope to discover a new group of fungi and would perhaps fill the space between known and unknown with some novel value. The secondary metabolites produced by fungal endophytes exhibited propitious results and can be considered as a contemporary substitute for obtaining potential drugs in the treatment of day by day emerging diseases. Albeit a lot of preliminary information is available pertaining to the bioactive metabolites, however, there is a colossal issue known as commercialization of the final product for the society is in front of the corporate and government organizations left to resolve. The key issue behind this problem was that, will the endophyte be able to produce the desired bioactive natural compounds continuously generation after generation. To deal with this predicament, in recent times workers from all over the world executed epigenetic modulation to induce the gene/or gene clusters through DNA modification, chromatin remodeling and mi-RNA. These transformations may encourage the formation of various 'cryptic metabolites' and may also improve the production of desired compounds by many folds to lower the unavailability of required secondary metabolites which were not produced during normal conditions.

Investigating the unexplored ecological units and environment for microbes including fungi is a lucrative enterprise for understanding their biology as well as in making the most out of their novel genes with the help of technology. Lichen thallus supports many different microbes and exhibits multipartite associations and because of that it symbolizes a miniature ecosystem – a befitting candidate for such investigations. The study on endolichenic fungi in India is in its juvenile stage and a lot needs to be done, not only regarding diversity but also on the bioactive potential of secondary metabolites isolated from these endolichenic fungi. The above mentioned researches support the fact that endolichenic fungi are present in practically every lichen species that have been examined to date, and stands as imperative however, inadequately investigated branch in lichenology. The endolichenic fungal species investigated till date for the isolation of bioactive secondary metabolites belonged to a handful of geographical locations. The estimated global lichen diversity is about 20,000 (Feuerer and Hawksworth 2007) and only a small number of lichen species have been screened for harvesting the endolichenic fungi with the potential to offer bioactive metabolites, hence one can assume the magnitude of prospective lichen diversity which is waiting to be unveiled. A lot of work is needed to be done on the genetic, molecular, and population aspect of endolichenic fungi to get a better perspective of the interaction between lichen and endolichenic fungi, and also to understand the ecological role that endolichenic fungi, and their metabolites, have in the symbiosis and fortification of the lichen thallus.

Documentations have been found in support of the positive role of fungal endophytes in plant defense in biotic and abiotic stresses. Lichens which used to grow in extreme environmental conditions can act as a model organism for carrying out studies on various abiotic stresses (thermal, metal and salt etc.) and herbivory by studying their endophytes. The exact molecular mechanism is required to know that whether endolichenic fungi are providing tolerance to lichen thallus against various stresses.

In recent past researchers have made a remarkable progress in culture-independent techniques and high-throughput sequencing methods and this has completely changed the face of the investigations on endolichenic fungi, but the phylogeny and taxonomy of most of these fungi have not been explored. This is because some of the endophytes are not culturable (Arnold et al. 2007; Pancher et al. 2012; Impullitti and Malvick 2013); some do not sporulate on artificial culture conditions and spores are significant structures in fungal taxonomy (Petrini and Petrini 1985); a majority of culture-independent methods depend on the nuclear ribosomal internal transcribed spacer region (nrITS), which is not open to broad-scale phylogenetic investigation, or use short reads that provide limited resolving power (Lindner and Banik 2011; Porter and Golding 2011). Besides all these limitations, the enormous amount of estimated diversity of endophytes is undocumented; thus, even if many sequences of endophytes have been deposited in GenBank, they usually provide limited taxonomic information (U'Ren et al. 2009; Gazis et al. 2012; Nilsson et al. 2014). Till the above said issues are not resolved, the taxonomic placement, ecological role and evolutionary details of endophytes cannot came into limelight.

Endolichenic fungi as a research subject are very intriguing, their every aspect has immense possibilities in answering the questions of basic science of lichen and apart from the academicians and researchers they have a huge pharmaceutical value which can fulfill the needs of the society.

Endophyte Definitions

Any organism occurring within the plant tissues (de Bary 1866).

Mutualists fungi that colonize aerial parts of living plant tissues and do not cause symptoms of disease (Carroll 1986).

Fungi that form unapparent infections within leaves and stems of healthy plants (Carroll 1988).

Fungi as colonizers of the living internal tissues of their plant host (Rollinger and Langenheim 1993).

All organisms inhabiting plant organs that at some time during their life can colonize internal plant tissues without causing apparent harm to the host (Petrini 1991).

A group that colonize living, internal tissues of plants without causing any immediate, overt negative effects (Hirsch and Braun 1992).

Any fungi isolated from internal symptomless plant tissues (Cabral et al. 1993).

Fungi and bacteria which, for all or part of their life cycle, invade the tissues of living plants and cause unapparent and asymptomatic infections entirely within plant tissues, but cause no symptoms of disease (Wilson 1995).

All the microbes that colonize living, internal tissues of plant without causing any immediate, overt negative effect (Bacon and White 2000).

True endophytes are those fungi whose colonization never results in visible disease symptoms (Mostert et al. 2000).

Fungi that colonize a plant without causing visible disease symptoms at any specific moment (Schulz and Boyle 2005).

Why are endophytes so alluring?

There are some logical justifications to study endophytes: (1) endophytes are highly diverse, can be worked out in laboratory with the minimal field-work required, and using a well-established traditional methodology, (2) diversity of endophytes mostly revolve around at least 50 genera, so it is easy for identification purposes, (3) various methodologies can be applied to mycelia sterilia to promote sporulation; alternatively molecular methods can be utilized to identify these relatively fast growing morphotypes, (4) statistical treatment can be given to the data of fungal isolates derived from single random units and will satisfy the demands of any unforgiving non-fungal ecologist, (5) the relatively fast growing and highly diverse endophytes provide ideal tools for screening and novel compounds discovery and they can easily be lodged in culture collections.

Difference between endophyte and mycorrhizae

Endophytic fungi differ from the mycorrhizal fungi in both ecology and infection allocation. While mycorrhizas are restricted to the rhizosphere with particular importance for plant growth in nutrient-stressed situations, endophytes can be found in both above and belowground plant-tissues, and seem to be present even when nutrients are abundant (Smith and Read 2008).

Difference between endolichenic and lichenicolous fungi

Endolichenic fungi differ from the lichenicolous fungi in both symptoms and phylogeny. Lichenicolous fungi show outward symptoms of infection and endolichenics lack any obvious outward symptoms. As far as phylogeny is considered, endolichenic fungi were found to be most common among primarily nonlichenized lineages of euascomycetes (Sordariomycetes, Dothideomycetes, Leotiomycetes and Pezizomycetes) but absent among the lichen-dominated clades (Lecanoromycetes, Arthoniomycetes, Lichinomycetes), while lichenicolous fungi were thought to be closely related to lichens and found mainly in lichen-forming clades (Arnold et al. 2009; Diederich et al. 2018). Besides this the commensalistic lichenicolous fungi would be nutritionally similar to primary mycobionts (Hawksworth 1988), while endolichenic fungi are associated mainly with the photobiont (Arnold et al. 2009).

References

Abreu-Tarazi MF, Navarrete AA, Andreote FD et al (2010) Endophytic bacteria in long-term in vitro cultivated axenic pineapple microplants revealed by PCR DGGE. World J Microbiol Biotechnol 26:555–560

Ahmadjian V (1993) The lichen symbiosis, 2nd edn. Wiley, New York

Arnold AE, Lutzoni F (2007) Diversity and host range of foliar fungal endophytes: are tropical leaves biodiversity hotspots? Ecology 88:541–549

Arnold AE, Mejía LC, Kyllo D et al (2003) Fungal endophytes limit pathogen damage in a tropical tree. Proc Natl Acad Sci U S A 100:15649–15654

Arnold AE, Henk DA, Eells RL et al (2007) Diversity and phylogenetic affinities of foliar fungal endophytes in loblolly pine inferred by culturing and environmental PCR. Mycologia 99:185–206

Arnold AE, Miadlikowska J, Higgins KL et al (2009) A phylogenetic estimation of trophic transition networks for ascomycetous fungi, are lichens cradles of symbiotrophic fungal diversification? Syst Biol 58:283–297

Bacon CW, White JF (2000) Microbial endophytes. Marcel Dekker Inc, New York

Bacon CW, Porter JK, Robbins JD et al (1977) *Epichloë typhina* from toxic tall fescue grasses. Appl Environ Microbiol 34:76–81

Bates ST, Cropsey GWG, Caporaso JG et al (2011) Bacterial communities associated with the lichen symbiosis. Appl Environ Microbiol 77:1309–1314

Bhardwaj A, Agrawal P (2014) A review fungal endophytes: as a storehouse of bioactive compound. World J Pharm Pharm Sci 3:228–237

Biosca EG, Flores R, Santander RD et al (2016) Innovative approaches using lichen enriched media to improve isolation and culturability of lichen associated bacteria. PLoS One 11(8):e0160328. https://doi.org/10.1371/journal.pone.0160328

Blackwell M (2011) The fungi: 1, 2, 3 … 5.1 million species? Am J Bot 98(3):426–438

Boonpragob K, Crittenden PD, Lumbsch TH (2012) Lichens: from genome to ecosystems in a changing world. MycoKeys 6:1–2

Cabral D, Stone J, Carroll GC (1993) The internal mycoflora of *Juncus* spp.: microscopic and cultural observations of infection patterns. Mycol Res 97:367–376

Cardinale M, de Castro JV, Müller H et al (2008) *In-situ* analysis of the bacterial community associated with the reindeer lichen *Cladonia arbuscula* reveals predominance of Alphaproteobacteria. FEMS Microbiol Ecol 66:63–71

Carroll G (1986) The biology of endophytism in plants with particular reference to woody perennials. In: Fokkema NJ, Van den Henves J (eds) Microbiology of the phyllosphere. Cambridge University Press, Cambridge, pp 205–222

Carroll GC (1988) Fungal endophytes in stems and leaves: from latent pathogen to mutualistic symbiont. Ecology 69:2–9

Carroll GC, Carroll F (1978) Studies on the incidence of coniferous needle endophytes in the Pacific northwest. Can J Bot 56:3034–3043

Carroll GC, Petrini O (1983) Patterns of substrate utilization by some endophytes from coniferous foliage. Mycologia 75:53–63

Castillo U, Strobel GA, Ford EJ et al (2002) Munumbicins, wide spectrum antibiotics produced by *Streptomyces munumbi*, endophytic on *Kennedia nigriscans*. Microbiology 148:2675–2685

Castillo U, Harper JK, Strobel GA et al (2003) Kakadumycins, novel antibiotics from *Streptomyces* sp. NRRL 30566, an endophyte of *Grevillea pteridifolia*. FEMS Microbiol Lett 224(2):183–190

Chagnon PL, U'Ren JM, Miadlikowska J et al (2016) Interaction type influences ecological network structure more than local abiotic conditions: evidence from endophytic and endolichenic fungi at a continental scale. Oecologia 180:181–191

Chambers SM, Williams PG, Seppelt RD et al (1999) Molecular identification of *Hymenoscyphus* sp. from the rhizoids of the leafy liverwort *Cephaloziella exiliflora* in Australia and Antarctica. Mycol Res 103:286–288

Chen KH, Miadlikowska J, Molnar K et al (2015) Phylogenetic analyses of eurotiomycetous endophytes reveal their close affinities to chaetothyriales, eurotiales, and a new order – phaeomoniellales. Mol Phylogenet Evol 85:117–130

Cheon DM, Jang DS, Kim HY et al (2013) Detection of antifungal endolichenic fungi and antifungal compound. Korean J Microbiol 49(2):165–171

Clay K (1991) Fungal endophytes, grasses and herbivores. In: Barbosa P, Krischik VA, Jones CG (eds) Microbial mediation of plant herbivore interactions. Wiley, New York, pp 199–226

Clay K, Schardl CL (2002) Evolutionary origins and ecological consequences of endophyte symbiosis with grasses. Am Nat 160:S99–S127

Crittenden PD, David JC, Hawksworth DL et al (1995) Attempted isolation and success in the culturing of a broad spectrum of lichen-forming and lichenicolous fungi. New Phytol 130:267–297

Cubit JD (1974) Interactions of seasonally changing physical factors and grazing affecting high intertidal communities on a rocky shore. PhD dissertation, University of Oregon

Davis EC, Franklin JB, Shaw AJ et al (2003) Endophytic *Xylaria* (Xylariaceae) among liverworts and angiosperms: phylogenetics, distribution, and symbiosis. Am J Bot 90:1661–1667

de Bary A (1866) Morphologie und Physiologie der Plize, Flechten, und Myxomyceten. In: Holfmeister's handbook of physiological botany, vol 2, Leipzig

Diederich P, Lawrey JD, Ertz D (2018) The 2018 classification and checklist of lichenicolous fungi, with 2000 non-lichenized, obligately lichenicolous taxa. The Bryologist 121(3):340–425

Ding G, Li Y, Fu S et al (2009) Ambuic acid and torreyanic acid derivatives from the endolichenic fungus *Pestalotiopsis* sp. J Nat Prod 72(1):182–186

Döbbler P (1979) Untersuchungen an moosparasitischen Pezizales aus der Verwandtschaft von *Octospora*. Nova Hedw 31:817–864

Elmi AA, West CP (1995) Endophyte infection effects on stomatal conductance, osmotic adjustment and drought recovery of tall fescue. New Phytol 131:61–67

Ezra D, Hess WH, Strobel GA (2004) New endophytic isolates of *M. albus*, a volatile antibiotic-producing fungus. Microbiology 150:4023–4031

Fernández-Mendoza A, Kopun FT, Grube M et al (2017) ITS1 metabarcoding highlights low specificity of lichen mycobiomes at a local scale. Mol Ecol 26:4811–4830

Feüerer T, Hawksworth DL (2007) Biodiversity of lichens, including a world-wide analysis of checklist data based on Takhtajan's floristic regions. Biodivers Conserv 16:85–98

Fisher PJ, Petrini O, Lappin Scott HM (1992) The distribution of some fungal and bacterial endophytes in maize (*Zea mays* L.). New Phytol 122:299–305

Fisher PJ, Graf F, Petrini LE et al (1995) Fungal endophytes of *Dryas octopetala* from a high polar semi desert and from the Swiss Alps. Mycologia 87:319–323

Fleischhacker A, Grube M, Kopun T et al (2015) Community analyses uncover high diversity of lichenicolous fungi in alpine habitat. Micro Ecol 70:348–360

Freeman EM (1904) The seed fungus of *Lolium temulentum* L., the darnel. Philos Trans R Soc Lond B 196:1–27

Fröhlich J, Hyde KD, Petrini O (2000) Endophytic fungi associated with palms. Mycol Res 104:1202–1212

Gams W (1971) *Cephalosporium*- artige Schimmelpilze (Hyphomycetes). Gustav Fisher Verlag, Stuttgard, pp 262

Gazis R, Miadlikowska J, Lutzoni F et al (2012) Culture-based study of endophytes associated with rubber trees in Peru reveals a new class of Pezizomycotina: Xylonomycetes. Mol Phylogenet Evol 65:294–304

Gerson U, Seaward MRD (1977) Lichen-invertebrate associations. In: Seaward MRD (ed) Lichen ecology. Academic, London, pp 69–120

Girlanda M, Isocrono D, Bianco C et al (1997) Two foliose lichens as microfungal ecological niches. Mycologia 89:531–536

Gond SK, Verma VC, Kumar A et al (2007) Study of endophytic fungal community from different parts of *Aegle marmelos* Correae (Rutaceae) from Varanasi (India). World J Microbiol Biotechnol 23:1371–1375

Grube M, Cernava T, Soh J et al (2015) Exploring functional contexts of symbiotic sustain within lichen-associated bacteria by comparative omics. ISME J 9:412–424
Guerin P (1898) Surla presence d'un chamignondansl'ivraie. J Botanique 12:230–238
Gunatilaka AAL (2006) Natural products from plant-associated microorganisms: distribution, structural diversity, bioactivity and implication of their occurrence. J Nat Prod 69:509–526
Hardoim P, van-Overbeek L, van-Elsas JD (2008) Properties of bacterial endophytes and their proposed role in plant growth. Trends Microbiol 16(10):463–471
Hawksworth DL (1975) Notes on British lichenicolous fungi. I Kew Bull 30:183–203
Hawksworth DL (1979) The lichenicolous Hyphomycetes. Bull Brit Mus (Nat Hist) Bot 6:183–300
Hawksworth DL (1980) Notes on some fungi occurring on *Peltigera*, with a key to accepted species. Trans Brit Mycol Soc 74:363–386
Hawksworth DL (1981) The lichenicolous Coelomycetes. Bull Brit Mus (Nat Hist) Bot 9:1–98
Hawksworth DL (1982) Secondary fungi in lichen symbioses: parasites, saprophytes and parasymbionts. J Hattori Bot Lab 52:357–366
Hawksworth DL (1983) A key to the lichen-forming, parasitic, parasymbiontic and saprophytic fungi occurring on lichens in the British Isles. Lichenologist 15:1–44
Hawksworth DL (1988) Effects of algae and lichen-forming fungi on tropical trees. In: Agnihotri VP, Sarbhoy AK, Kumar D (eds) Perspectives in mycology and plant pathology. Malhotra, New Delhi, pp 76–83
Hawksworth DL, Jones D (1981) *Sclerococcum sphaerale* obtained in pure culture. Trans Br Mycol Soc 77:485–489
Hawksworth DL, Rossman AY (1997) Where are all the undescribed fungi? Phytopathology 87:888–891
Heckman DS, Geiser DM, Eidell BR et al (2001) Molecular evidence for the early colonization of land by fungi and plants. Science 293(5532):1129–1133
Hibbett DS, Taylor JW (2013) Fungal systematics: is a new age of enlightenment at hand? Nat Rev Microbiol 11(2):129–133
Higgins KL, Arnold AE, Miadlikowska J et al (2007) Phylogenetic relationships, host affinity, and geographic structure of boreal and arctic endophytes from three major plant lineages. Mole Phylogenet Evol 42:543–555
Hirsch GU, Braun U (1992) Communities of parasitic microfungi. In: Winterholf W (ed) Handbook of vegetation science: fungi in vegetation science, vol 19. Kluner Academic, Dordrecht, pp 225–250
Hodkinson BP, Lutzoni F (2009) A microbiotic survey of lichen-associated bacteria reveals a new lineage from the Rhizobiales. Symbiosis 49:163–180
Hofstetter V, Miadlikowska J, Kauff F et al (2007) Phylogenetic comparison of protein-coding versus ribosomal RNA-coding sequence data: a case study of the Lecanoromycetes (Ascomycota). Mol Phylogenet Evol 44:412–426
Honegger R (1996) Mycobionts. In: Nash TH III (ed) Lichen biology. Cambridge University Press, Cambridge, pp 24–36
Hwang HG, Kim YN, Baik KS et al (2011) Isolation and characteristics of endolichenic fungi producing antifungal compound. Korean J Microbiol 47(1):97–101
Impullitti AE, Malvick DK (2013) Fungal endophyte diversity in soybean. J Appl Microbiol 114:1500–1506
Jalgaonwala RE, Mohite BV, Mahajan RT (2011) Natural products from plant associated endophytic fungi. J Microbiol Biotechnol Res 1:21–32
Janardhanan KK, Ahmad A (1997) Fungal endophytes of grasses: incidence and distribution in India. In: Janardhanan KK, Natarajan K, Hawksworth DL (eds) Tropical mycology. Oxford and IBH Publishing Co Pvt. Ltd, Calcutta, pp 157–168
Jayakumar K, Prasad Babu G, Charan Theja P et al (2016) Biomedical potentials of *Talaromyces tratensis*– a new endolichenic fungi associated with high altitude curstose lichen *Lecanora* sp. Res J Pharm Biol Chem Sci 7(5):1412–1419
Jeffries P, Young TWK (1994) Interfungal parasitic relationships. CAB International, Wallingford, pp 296

Johnson JA, Whitney NJ (1992) Isolation of fungal endophytes from black spruce (*Picea mariana*) dormant buds and needles from New Brunswick, Canada. Can J Bot 70(9):1754–1757

Kannangara BT, Rajapaksha RS, Paranagama PA (2009) Nature and bioactivities of endolichenic fungi in *Pseudocyphellaria* sp., *Parmotrema* sp. and *Usnea* sp. at Hakgala montane forest in Sri Lanka. Lett Appl Microbiol 48:203–209

Kauserud H, Mathiesen C, Ohlson M (2008) High diversity of fungi associated with living parts of boreal forest bryophytes. Botany 86:1326–1333

Kharwar RN, Verma VC, Strobel G et al (2008) The endophytic fungal complex of *Catharanthus roseus* (L.) G. Don. Curr Sci 95:228–233

Kharwar RN, Gond SK, Kumar A et al (2010) A comparative study of endophytic and epiphytic fungal association with leaf of *Eucalyptus citriodora* hook., and their antimicrobial activity. World J Microbiol Biotechnol 26:1941–1948

Kharwar RN, Mishra A, Gond SK et al (2011) Anticancer compounds derived from fungal endophytes: their importance and future challenges. Nat Prod Rep 28:1208–1228

Kim ES, Choi KS, Choi SK (2012) Detection of endolichenic fungi producing antifungal compound. Korean J Microbiol Biotechnol 40(1):23–29

Kirk PM, Cannon PF, Minter DW et al (2008) Dictionary of the fungi, 10th edn. CABI, Wallingford

Krings M, Taylor TN, Hass H et al (2007) Fungal endophytes in a 400-million-yr-old land plant: infection pathways, spatial distribution, and host responses. New Phytol 174(3):648–657

Kröhn K, Beckmann K, Flörke U et al (1997) Biologically active metabolites from fungi, 9 new palmarumycins CP4a und CP5 from *Coniothyrium palmarum*: structure elucidation, crystal structure analysis and determination of the absolute configuration by CD-calculations. Tetrahedron 53:3101–3110

Kröhn K, Biele C, Drogies KH et al (2002) Biologically active secondary metabolites from fungi, 18 fusidilactones, a new group of polycyclic lactones from an endophyte, *Fusidium* sp. Eur J Org Chem 2002:2331–2334

Kröhn K, Dai J, Florke U et al (2005) Botryane metabolites from the fungus *Geniculosporium* sp. isolated from the marine red alga *Polysiphonia*. J Nat Prod 68:400–405

Kumaresan V, Ganesan T, Rajarajan D et al (2006) Fungal endophytes of *Psilotum nudum*—a first report. Geobios 33:200–202

Kumaresan V, Veeramohan R, Bhat MM et al (2013) Fungal endophyte assemblages of some Pteridophytes from Mahe, India. World J Sci Technol 3(1):7–10

Lawrey JD (1995) The chemical ecology of lichen mycoparasites. Can J Bot 73:603–608

Lawrey JD, Diederich P (2003) Lichenicolous fungi: interactions, evolution, and biodiversity. The Bryologist 106:80–120

Li JY, Sidhu RS, Ford EJ et al (1998) The induction of taxol production in the endophytic fungus (*Periconia* sp.) from *Torreya grandifolia*. J Ind Microbiol Biotechnol 20:259–264

Li HY, Qing C, Zhang YL et al (2005) Screening for endophytic fungi with antitumor and antifungal activities from Chinese medicinal plants. World J Microbiol Biotechnol 21:1515–1519

Li WC, Zhou J, Guo SY et al (2007) Endophytic fungi associated with lichens in Baihua mountain of Beijing, China. Fungal Divers 25:69–80

Ligrone R (1988) Ultrastructure of a fungal endophyte in *Phaeoceros laevis* (L.) Prosk, anthocerotophyta. Bot Gaz 149(1):92–100

Ligrone R, Pocock K, Duckett JG (1993) A comparative ultrastructural study of endophytic basidiomycetes in the parasitic achlorophyllous hepatic *Cryptothallus mirabilis* and the closely allied photosynthetic species *Aneura pinguis* (Metzgeriales). Can J Bot 71:666–679

Lin X, Lu C, Huang Y et al (2007) Endophytic fungi from a pharmaceutical plant, *Camptotheca acuminata*: isolation, identification and bioactivity. World J Microbiol Biotechnol 23:1037–1040

Lindner DL, Banik MT (2011) Intragenomic variation in the ITS rDNA region obscures phylogenetic relationships and inflates estimates of operational taxonomic units in genus *Laetiporus*. Mycologia 103:731–740

Lodge JD, Fisher PJ, Sutton BC (1996) Endophytic fungi of *Manikara bidentata* leaves in Puerto Rico. Mycologia 88(5):733–738

Lutzoni F, Pagel M, Reeb V (2001) Major fungal lineages are derived from lichen symbiotic ancestors. Nature 411:937–940

Maduranga K, Attanayake RN, Santhirasegaram S et al (2018) Molecular phylogeny and bioprospecting of Endolichenic fungi (ELF) inhabiting in the lichens collected from a mangrove ecosystem in Sri Lanka. PLoS One 13(8):e0200711. https://doi.org/10.1371/journal.pone.0200711

Malinowski DP, Zuo H, Belesky DP et al (2004) Evidence for copper binding by extracellular root exudates of tall fescue but not perennial ryegrass infected with *Neotyphodium* spp. endophytes. Plant Soil 267:1–12

Mathan S, Subramanian V, Nagamony S et al (2013) Isolation of endophytic fungi from marine algae and bioactivity. Int J Res Pharm Sci 4(1):45–49

Miadlikowska A, Arnold A, Lutzoni F (2004) Diversity of cryptic fungi inhabiting healthy lichen thalli in a temperate and tropical forest. Ecol Soc Am Annu Meet 89:349–350

Mishra Y, Singh A, Batra A et al (2014) Understanding the biodiversity and biological applications of endophytic fungi: a review. J Microbial Biochem Technol S8 https://doi.org/10.4172/1948-5948.S8-004

Mostert L, Crous PW, Petrini O (2000) Endophytic fungi associated with shoots and leaves of *Vitis vinifera*, with specific reference to the *Phomopsis viticola* complex. Sydowia 52:46–58

Muggia L, Fleischhacker A, Kopun T et al (2016) Extremotolerant fungi from alpine rock lichens and their phylogenetic relationships. Fungal Divers 76:119–142

Mushin TM, Booth T (1987) Fungi associated with halophytes of an inland salt marsh, Manitoba, Canada. Can J Bot 65:1137–1151

Mushin TM, Booth T, Zwain KH (1989) A fungal endophyte associated with a desert parasitic plant. Kavaka 17:1–5

Nalini MS, Mahesh B, Tejesvi MV et al (2005) Fungal endophytes from the three-leaved caper, *Crataeva magna* (Lour.) DC. (Capparidaceae). Mycopathologia 159:245–249

Nilsson RH, Hyde KD, Pawłowska J et al (2014) Improving ITS sequence data for identification of plant pathogenic fungi. Fungal Divers 67:11–19

Padhi S, Tayung K (2015) *in-vitro* antimicrobial potentials of endolichenic fungi isolated from thalli of *Parmelia* lichen against some human pathogens. Beni Suef Univ J Basic Appl Sci 4(4):299–306

Pancher M, Ceol M, Corneo PE et al (2012) Fungal endophytic communities in grapevines (*Vitis vinifera* L.) respond to crop management. Appl Environ Microbiol 78:4308–4317

Paranagama PA, Wijeratne EMK, Burns AM et al (2007) Heptaketides from *Corynespora* sp. inhabiting the cavern beard lichen, *Usnea cavernosa*: first report of metabolites of an endolichenic fungus. J Nat Prod 70(11):1700–1705

Park JH, Choi GJ, Lee HB et al (2005) Griseofulvin from *Xylaria* sp. strain F0010, an endophytic fungus of *Abies holophylla* and its antifungal activity against plant pathogenic fungi. J Microbiol Biotechnol 15(1):112–117

Park CH, Kim KM, Elvebakk A et al (2014) Algal and fungal diversity in Antarctic lichens. J Eukaryot Microbiol 62:196–205

Petrini O (1987) Endophytic fungi of alpine Ericaceae. In: Laursen GA, Ammirati JF, Redhead SA (eds) The endophytes of *Loiseleuria procumbens*. Arctic and alpine mycology II. Plenum, New York, pp 71–77

Petrini O (1991) Fungal endophytes of tree leaves. In: Andrews J, Hirano S (eds) Microbial ecology of leaves. Springer, New York, pp 179–197

Petrini LE, Petrini O (1985) Xylariaceous fungi as endophytes. Sydowia 38:216–234

Petrini O, Hake U, Dreyfuss M (1990) An analysis of fungal communities isolated from fruticose lichens. Mycologia 82:444–451

Petrini O, Sieber T, Toti L et al (1992) Ecology, metabolite production, and substrate utilization in endophytic fungi. Nat Toxins 1:185–196

Pocock K, Duckett JG (1985) Fungi in hepatics. Bryol Times 31:2–3

Porras-Alfaro A, Bayman P (2011) Hidden fungi, emergent properties: endophytes and microbiomes. Annu Rev Phytopathol 49:291–315

Porter TM, Golding GB (2011) Are similarity or phylogeny-based methods more appropriate for classifying internal transcribed spacer (ITS) metagenomic amplicons? New Phytol 192:775–782
Rajagopal R, Suryanarayanan TS (2000) Isolation of endophytic fungi from leaves of neem (*Azadirachta indica*). Curr Sci 78:1375–1378
Raviraja NS (2005) Fungal endophytes in five medicinal plant species from Kudremukh range, Western Ghats of India. J Basic Microbiol 45:230–235
Redman RS, Kim YO, Woodward CJDA et al (2011) Increased fitness of rice plants to abiotic stress via habitat adapted symbiosis: a strategy for mitigating impacts of climate change. PLoS One 6(7):e14823. https://doi.org/10.1371/journal.pone.0014823
Remy W, Taylor TN, Hass H et al (1994) 4-hundred-million-year-old vesicular arbuscular mycorrhizae. Proc Natl Acad Sci U S A 91(25):11841–11843
Riyaz-Ul-Hassan S, Strobel GA, Booth E et al (2012) Modulation of volatile organic compound formation in the mycodiesel producing endophyte-*Hypoxylon* sp. C1-4. Microbiology 158:464–473
Rodriguez RJ, White JF, Arnold AE et al (2009) Fungal endophytes: diversity and functional roles. New Phytol 182(2):314–330
Rollinger J, Langenheim JH (1993) Geographic survey in fungal endophyte community composition in leaves of redwood. Mycologia 85:149–156
Rosenblueth M, Martínez-Romero E (2006) Bacterial endophytes and their interactions with hosts. Mol Plant-Microbe Interact 19(8):827–837
Sahashi N, Kubono T, Miyasawa Y et al (1999) Temporal variations in isolation frequency of endophytic fungi from Japanese beech. Can J Bot 77:197–202
Saikkonen K, Faeth SH, Helander M et al (1998) Fungal endophytes, a continuum of interactions with host plants. Ann Rev Ecol Syst 29:319–343
Sati SC, Belwal M (2005) Aquatic hyphomycetes as endophytes of riparian plant roots. Mycologia 97(1):45–49
Sati SC, Pargaein N, Belwal M (2009) Diversity of aquatic hyphomycetes as root endophytes on pteridophytic plants in Kumaun Himalaya. J Am Sci 5(4):179–182
Schardl CL, Leuchtmann A, Spiering MJ (2004) Symbioses of grasses with seedborne fungal endophytes. Ann Rev Plant Biol 55:315–340
Schmid E, Oberwinkler F (1993) Mycorrhiza-like interaction between the achlorophyllous gametophyte of *Lycopodium clavatum* L. and its fungal endophyte studied by light and electron microscopy. New Phytol 124:69–81
Schulz B, Boyle C (2005) The Endophytic continuum. Mycol Res 109:661–686
Schulz B, Boyle C, Draeger S et al (2002) Endophytic fungi: a source of novel biologically active secondary metabolites. Mycol Res 106:996–1004
Selosse MA, Le Tacon F (1998) The land flora: a phototroph-fungus partnership? Trend Ecol Evol 13(1):15–20
Shanmugam P, Ponnusamy P, Balasubramanian G et al (2018) Screening of biologically potent endolichenic fungi isolated from selected lichens habitat on silver oak tree. Vegetos – An Int J Pl Res 31(3):89–94
Shipunov A, Newcombe G, Raghavendra AKH et al (2008) Hidden diversity of endophytic fungi in an invasive plant. Am J Bot 95(9):1096–1108
Sieber T (2007) Endophytic fungi in forest trees: are they mutualists? Fungal Biol Rev 21:75–89
Singh BN, Upreti DK, Gupta VK et al (2017) Endolichenic fungi: a hidden reservoir of next generation biopharmaceuticals. Trends Biotechnol 35(9):808–813
Smith SE, Read DJ (2008) Mycorrhizal symbiosis. Academic, London
Stierle A, Strobel G, Stierle D (1993) Taxol and taxane production by *Taxomyces andreanae*, an endophytic fungus of Pacific yew. Science 260:214–216
Stone JK, Bacon CW, White JF (2000) An overview of endophytic microbes: endophytism defined. In: Bacon CW, White JF (eds) Microbial Endophytes. Marcel Dekker, New York, pp 3–30

Stone JK, Polishook JD, White JF Jr (2004) Endophytic fungi. In: Mueller GM, Bills GF, Foster MS (eds) Biodiversity of fungi: inventory and monitoring methods. Elsevier Academic Press, New York, pp 241–270

Strobel G (2006) Harnessing endophytes for industrial microbiology. Curr Opin Microbiol 9:240–244

Strobel GA, Daisy B (2003) Bioprospecting for microbial endophytes and their natural products. Microbiol Mol Biol Rev 67:491–502

Strobel GA, Hess WM, Ford E et al (1996) Taxol from fungal endophytes and the issue of biodiversity. J Ind Microbiol 17:417–423

Suryanarayanan TS, Shaanker RU (2009) How promising are endophytic fungi as alternative sources of plant secondary metabolites? Curr Sci 97:477–478

Suryanarayanan TS, Thirunavukkarasu N, Hariharan GN et al (2005) Occurrence of non-obligate microfungi inside lichen thalli. Sydowia 57:120–130

Suryanarayanan TS, Govindarajulu MB, Rajamani T et al (2017) Endolichenic fungi in lichens of Champawat district, Uttarakhand, Northern India. Mycol Progress. https://doi.org/10.1007/s11557-016-1268-7

Suryanarayanana TS, Thirunavukkarasu N (2017) Endolichenic fungi: the lesser known fungal associates of lichens. Mycology 8(3):189–196

Swatzell LJ, Powell MJ, Kiss JZ (1996) The relationship of endophytic fungi to the gametophyte of the fern *Schizaea pusilla*. Int J Plant Sci 157:53–62

Sydowia H (1914) Beitragezurkenntnis der pilzflora des sudlichenostindiens II. Ann Mycologica 12:484–490

Tan RX, Zou WX (2001) Endophytes: a rich source of functional metabolites. Nat Prod Rep 18:448–459

Taylor JE, Hyde KD, Jones EBG (1999) Endophytic fungi associated with the temperate palm, *Trachycarpus fortunei*, within and outside its natural geographic range. New Phytol 142(2): 335–346

Tejesvi MV, Mahesh B, Nalini MS et al (2005) Endophytic fungal assemblages from inner bark and twig of *Terminalia arjuna* W. And A. (Combretaceae). World J Microbiol Biotechnol 21:1535–1540

Thirunavukkarasu N, Suryanarayanan TS, Murali TS et al (2011) L-asparaginase from marine derived fungal endophytes of seaweeds. Mycosphere 2(2):147–155

Thongsandee W, Matsuda Y, Ito S (2012) Temporal variations in endophytic fungal assemblages of *Ginkgo biloba* L. J For Res 17(2):213–218

Torzilli AP, Lawrey JD (1995) Lichen metabolites inhibit cell wall-degrading enzymes produced by the lichen parasite *Nectria parmeliae*. Mycologia 87:841–845

Tripathi M, Joshi Y (2015) Endolichenic fungi in Kumaun Himalaya: a case study. In: Upreti DK, Divakar PK, Shukla V et al (eds) Recent advances in lichenology. Springer, New Delhi, pp 111–120

Tripathi M, Gupta RC, Joshi Y (2014a) *Spegazzinia tessarthra* isolated as a true endophyte from lichen *Heterodermia flabellata*. Indian Phytopathol 67(1):109–110

Tripathi M, Gupta RC, Joshi Y (2014b) *Physcia dilatata* Nyl. (lichenized fungi, Physciaceae); a new host of *Bipolaris australiensis* (M.B. Ellis) Tsuda and Ueyama from Kumaun Himalaya, India. Proc Nat Acad Sci Lett 37(5):477–479

Tripathi M, Joshi Y, Gupta RC (2014c) Assessment of endolichenic fungal diversity in some forests of Kumaun Himalaya. Curr Sci 107(5):745–748

Tripathi M, Gupta RC, Joshi Y (2014d) Macrolichens as an excellent host of endophytic fungi: a case study in Banri Devi forest of Kumaun Himalaya, Uttarakhand, India. In: Jha P (ed) Biodiversity, conservation and sustainable development. New Academic Publishers, New Delhi, pp 141–146

U'Ren JM (2011) Host-geographic and ecological specificity of endophytic and endolichenic fungal communities. Ph.D. thesis, University of Arizona

U'Ren JM, Dalling JW, Gallery RE et al (2009) Diversity and evolutionary origins of fungi associated with seeds of a neotropical pioneer tree: a case study for analyzing fungal environmental samples. Mycol Res 113:432–449

U'Ren JM, Lutzoni F, Miadlikowska J et al (2010) Community analysis reveals close affinities between endophytic and endolichenic fungi in mosses and lichens. Microb Ecol 60:340–353

U'Ren JM, Lutzoni F, Miadlikowska J et al (2012) Host and geographic structure of endophytic and endolichenic fungi at a continental scale. Am J Bot 99:898–914

U'Ren JM, Riddle JM, Monacell JT et al (2014) Tissue storage and primer selection influence pyrosequencing-based inferences of diversity and community composition of endolichenic and endophytic fungi. Mol Ecol Resour 14:1032–1048

U'Ren JM, Miadlikowska J, Zimmerman NB et al (2016) Contributions of north American endophytes to the phylogeny, ecology, and taxonomy of Xylariaceae (Sordariomycetes, Ascomycota). Mol Phylogenet Evol 98:210–232

Verma VC, Gond SK, Kumar A et al (2007) Endophytic mycoflora from leaf, bark, and stem of *Azadirachta indica* Juss. from Varanasi India. Microb Ecol 54:119–125

Verma SK, Gond SK, Mishra A et al (2014) Impact of environmental variables on the isolation, diversity and antibacterial activity of endophytic fungal communities from *Madhuca indica* Gmel. At different locations in India. Ann Microbiol 64(2):721–734

Vinayaka KS, Krishnamurthy YL, Banakar S et al (2016) Association and variation of endophytic fungi among some macrolichens in Central Western Ghats, southern India. Int J Curr Microbiol App Sci 5(6):115–124

Wang S, Li XM, Teuscher F et al (2006) Chaetopyranin, a benzaldehyde derivative, and other related metabolites from *Chaetomium globosum*, an endophytic fungus derived from the marine red alga *Polysiphonia urceolata*. J Nat Prod 69:1622–1625

Wang FW, Jiao RH, Cheng AB et al (2007) Antimicrobial potentials of endophytic fungi residing in *Quercus variabilis* and brefeldin a obtained from *Cladosporium* sp. World J Microbiol Biotechnol 23(1):79–83

Wang Y, Zheng Y, Wang X et al (2016) Lichen-associated fungal community in *Hypogymnia hypotrypa* (Parmeliaceae, Ascomycota) affected by geographic distribution and altitude. Front Microbiol 7:1231. https://doi.org/10.3389/fmicb.2016.01231

Wilson D (1995) Endophyte – the evolution of a term, and clarification of its use and definition. Oikos 73:274–276

Yang RY, Li CY, Lin YC et al (2006) Lactones from a brown alga endophytic fungus (no. ZZF36) from the South China Sea and their antimicrobial activities. Bioorganic Med Chem Lett 16:4205–4208

You F, Han T, Wu J et al (2009) Antifungal secondary metabolites from endophytic *Verticillium* sp. Biochem Syst Ecol 37(3):162–165

Zhang HW, Song YC, Tan RX (2006) Biology and chemistry of endophytes. Nat Prod Rep 23:753–771

Zhang T, Zhang YQ, Liu HY et al (2013) Diversity and cold adaptation of culturable endophytic fungi from bryophytes in the fields region, King George island, maritime Antarctica. FEMS Microbiol Lett 341(1):52–61

Zimmerman NB, Vitousek PM (2012) Fungal endophyte communities reflect environmental structuring across a Hawaiian landscape. Proc Natl Acad Sci U S A 109(32):13022–13027

Methods for Isolation of Endolichenic Fungi and their Secondary Metabolites

Abstract Lichens are the ecological playground for many fungi, which can be categorized on the basis of their role inside the thallus, e.g. fungus which is in true symbiotic association with the algal partner, called mycobiont; fungi which are living inside the thallus in asymptomatic hyphal form and do not produce any reproductive bodies, called endolichenic fungi; fungi which cause disease on the thallus and produce reproductive bodies, lichenicolous fungi. In this book we are discussing about the endolichenic fungi, so this chapter is dedicated to the methods for isolating, culturing and preserving the endolichenic fungi and the secondary metabolites produced by them.

Lichens in general are comprised of a symbiotic association of two permanent members [mycobiont (primary fungus) and phycobiont/cyanobiont] and in particular some temporary associates [endolichenic and lichenicolous fungi (secondary fungi)] making this relationship somewhat complex to understand. Hence, to understand the interactions of these different organisms and the nature of symbiosis in lichens one has to separate, isolate and culture each of them first. There are several publications and even books available on the culture of mycobionts, photobionts, lichen thalli and lichenicolous fungi, but there has been scanty literature dedicated to the protocols for isolating the endolichenic fungi. The isolation of endolichenic fungi is necessary to establish their relationship with lichens and to solve the mystery of their existence inside the lichen thallus and their status of symbiosis. These endolichenic fungi are generally very tricky to culture because the hyphae of these fungi are found intertwined with both mycobiont and phycobiont and the surface of the lichens are often contaminated with bacteria, foreign algae and fungi. To remove these contaminants there are some surface sterilization protocols available which are necessary to be followed (Table 1).

The objective of this chapter is to expound the protocols for the isolation, culture and preservation of endolichenic fungi from lichen thalli. This chapter explains various alternative techniques for surface sterilization of lichen thalli and provides

M. Tripathi, Y. Joshi, *Endolichenic Fungi: Present and Future Trends*,
https://doi.org/10.1007/978-981-13-7268-1_3

Table 1 Different surface sterilization protocols used to clean the lichen thalli

S. No.	Surface sterilization protocols (SSP)	Reference
1	Lichen sample was taken in a tube with 30 ml sterile tap water (STW) and shaken on a rotary-shaker for 5 min at 200 rpm and after that water was discarded. The step was repeated once more to remove as much superficial contaminants as possible. Finally 20 ml STW was added to the washed material, homogenized with sterile bar blender-units (Janke & Kunkel) and the homogenate was passed first through a 125 μm and then a 40 μm mesh sieve. Lichen fragments held back in the 40 μm mesh were re-suspended in ca. 5 ml STW. One to five drops of the resulting suspension were deposited on the agar surface of the isolation plates and dispersed evenly by flooding with 1 ml STW.	Petrini et al. (1990)
2	Lichen samples were serially washed 20 times in sterile water (90 s per wash) under agitation followed by immersing the segments in 30% H_2O_2 (90 s) and finally washed four times (15 min each) in sterile water.	Girlanda et al. (1997)
3	SSP I: Segments were merely washed in sterile water. SSP II: 20 serial washes in sterile water (90 s per wash) under agitation. SSP III: 20 serial washes in sterile water (90 s per wash) under agitation followed by immersing the segments in 30% H_2O_2 (90 s) and finally washed four times (15 min each) in sterile water. SSP IV: The segments were dipped in 70% ethanol for 5 s, followed by 4% NaOCl for 90 s and sterile water for 10 s	Suryanarayanan et al. (2005)
4	Lichen thalli were cleaned in tap water to remove excess earth and litter. Then they were thoroughly washed under running tap water and then surface sterilized by consecutively immersing them in 75% ethanol (1 min), 2% NaOCl (3 min) and 75% ethanol (30 s). The sterilized thalli were surface dried with sterile paper towels and cut into segments of ca. 0.5 × 0.5 cm.	Li et al. (2007)
5	Lichen samples were washed thoroughly in running tap water to remove the debris. Then the samples were dipped in petri plate containing sterile double distilled water and bryophytes/mosses were removed and samples were then again washed in sterile double distilled water 20 times until all the other visible contaminants get removed. The samples were then subjected to chemical surface sterilization by dipping them in 30% hydrogen peroxide (H_2O_2) for 30 s, followed by 4% sodium hypochlorite (NaOCl) for 30 s and finally immersing them in 75% ethanol for 30 s. After chemical surface sterilization the samples were rinsed in double distilled autoclaved water twice and dried under aseptic conditions and were cut into small pieces (0.5 × 0.5 cm).	Tripathi and Joshi (2015)

various alternative methods for culturing and preserving them. These methodologies have been reported by various workers throughout the globe, but we have compiled them here in a single chapter for the ease of the researchers working in the field of endophytes and also added some general strategies to be followed while culturing them (Table 2).

Table 2 Different culture protocols followed by various workers

Parameters	Petrini et al. (1990)	Girlanda et al. (1997)	Suryanarayanan et al. (2005)	Li et al. (2007)	Tripathi and Joshi (2015)
Habit	Fruticose	Foliose	Foliose and fruticose	Crustose, foliose and fruticose	Foliose and fruticose
Habitat	Terricolous	Saxicolous and Terricolous	Corticolous	Corticolous, Saxicolous and Terricolous	Corticolous, Saxicolous and Terricolous
Sample processing	Within 24 h of collection	Not mentioned	Within 5 h of collection	Within 7 days of collection	Within 24 h of collection
Segment size	Homogenized thallus	0.25 cm^2	0.50 cm^2	0.50 cm^2	Random sized or 0.25 cm^2
Media	Malt extract agar with yeast extract	Malt extract agar with yeast extract	Potato dextrose agar	Malt extract agar	Potato dextrose agar
Antibiotics	Chlor-tetracycline	None	Chloramphenicol	Streptomycin sulphate	Streptomycin
Fungal growth inhibitor	Cyclosporin A	None	None	Rose Bengal	None
Incubation temperature	21 °C	Room temperature	26 °C	25 °C	25 °C
Incubation duration	3 weeks	Not mentioned	4 weeks	8 weeks	4 weeks
Media used for sub-culturing	Malt extract agar	Malt extract agar	Potato dextrose agar	Malt extract agar	Potato dextrose agar, malt extract agar and yeast malt extract agar

The process of isolation of endolichenic fungi broadly involves collection of lichen sample, sterilizing the lichen thallus and emergence of endolichenic fungi. To get the desired results all these steps are described here in detail (Fig. 1).

1 Lichen Collection

In comparison with higher plants here in case of lichens the sample often comes along with its substrate (bark, rock, soil or leaf) and debris, so there is a possibility for changes in diversity of endolichenic fungal colonies if the sample is not processed quickly. However, to get positive results, the collected lichen samples should be processed as soon as possible, generally within 2 days of collection. The samples

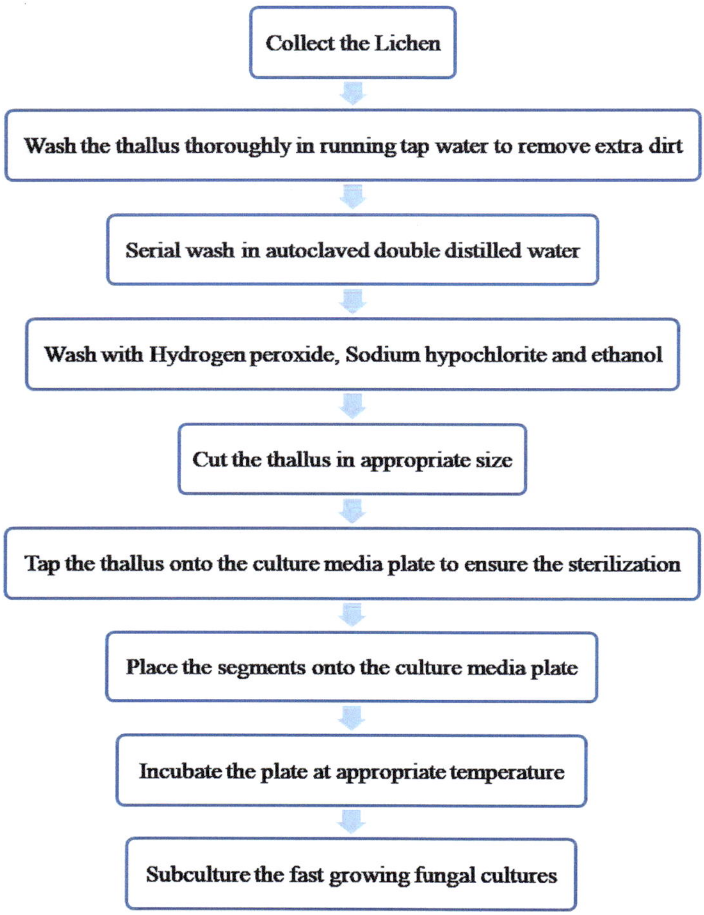

Fig. 1 Stepwise methodology to be followed for the isolation of endolichenic fungi

should be collected in sterilized paper bags instead of plastic bags so as to avoid moisture accretion followed by growth of superficial molds. The bag containing samples must bear field label comprising certain information about the sample including: on field identification or provisional name of the lichen sample, name of the collector, collection number, date, detailed information of the locality, including latitudes and longitudes and altitude, and habitat descriptions, substratum and host. Field notes should be deposited with the samples to smooth the progress of their reclamation and characterization.

2 Surface Sterilization

The first and foremost step in the isolation procedure of endolichenic fungi is to sterilize the surface of lichen thallus and this may vary according to the habit of lichen (e.g. crustose, foliose or fruticose), as crustose and some foliose lichens do not contain the lower cortex while fruticose and other foliose lichens bears cortex on both sides. Similarly, lichens growing on rocks (saxicolous) and soils (terricolous) require more rigorous cleaning in comparison with those which are growing on bark of trees (corticolous). Hence, before performing any detailed investigation the investigators should optimize the specific surface sterilization protocol according to the target lichen species (i.e. habit and habitat) for the recovery of maximum number of endolichenic fungi from that particular species.

Surface sterilization of lichen thallus typically includes serial rinsing with sterile water to remove extra dirt or debris, followed by treatment with a strong oxidant (H_2O_2) and general disinfectant (NaOCl) for a brief period which varies with lichen species. The dilution and duration of exposure to oxidants and disinfectants should be specified accordingly. The use of these sterilants for surface sterilization in case of lichens started from 1997, when Girlanda et al. (1997) used 30% H_2O_2 as a chemical surface sterilant. Suryanarayanan et al. (2005) evaluated the potential of sodium hypochlorite and ethanol in the sterilization process. Li et al. (2007) and Tripathi and Joshi (2015) used all the above discussed chemicals and serial washed the subjected lichen thallus. However, of the so far known surface sterilization protocols for endolichenic fungi, the protocol given by Petrini et al. (1990) was the only one which did not use any chemicals during sterilization procedure. Hence from 1997 onwards the use of chemical sterilants during the surface sterilization procedure of lichen thallus came into existence and is increasing steadily. In most of the cases ethanol came out as the most common surface sterilant as it gives the positive synergistic effect with other sterilants. However, one should keep in mind that, ethanol possesses restricted antibiotic potential and hence, should not be used alone as a surface disinfectant (Schulz et al. 1993).

After following all the steps of surface sterilization procedure the efficiency of the sterilizing agents should also be checked before by placing the thallus segments on to the culture plate. For that purpose surface sterilized segments of lichen thallus should be tapped on to the culture plates and allowed to incubate; if no growth is found on the culture plate then the thallus segments are perfectly sterilized and are ready to be incubated (Schulz et al. 1993). Besides this, the remains of the final washing can also be used to evaluate the sterilization procedure. Four to five drops of the remains of washing should be spread onto the culture plate and allowed to incubate and then checked for any growth.

3 Cutting the Lichen Thallus

Cutting lichen thallus tissue into many small pieces is quite laborious and varies with the growth form of lichens. Foliose and fruticose lichens can be cut by a scissor while crustose lichens can be scraped off from the substrate with the help of a sharp razor blade. Alternatively, washed and disinfected lichen thallus can be homogenized with the help of sterile bar-blender units. The lichen thallus gets macerated proficiently into small fragments which are suitable for direct dilution plating. The homogenate is passed through mesh sieve and the drops of the resulting suspension should be deposited on the agar surface of the isolation plates and dispersed evenly by flooding with sterile tap water (Petrini et al. 1990).

4 Culture of Endolichenic Fungi

Media Routinely used mycological media are appropriate for the primary isolation of endolichenic fungi and to subculture them for identification purpose. Malt extract agar and Potato dextrose agar are used generally, but sometimes Malt extract agar is used in combination with yeast extract.

Antibiotics and Growth Limiting Agents Antibiotics and fungal colony growth limiting agents are also used during the primary isolation. In case of antibiotics there is no common antibiotic agent found to be used, and its choice is quite random among all the published protocols. Petrini et al. (1990) and Li et al. (2007) used Cyclosporin A and Rose Bengal as fungal growth limiting agents, respectively, while Girlanda et al. (1997), Suryanarayanan et al. (2005) and Tripathi and Joshi (2015) didn't use any fungal growth inhibitors in their study.

Incubation Endolichenic fungi are very sluggish to appear, hence, sometimes the incubation period has to be protracted, because of which media may gets desiccated. To avoid desiccation of media, the culture plates should be sealed with Parafilm and humidity should be maintained in the growth chamber. The effects of light and dark on the emergence of endolichenic fungi has been tested by Suryanarayanan et al. (2005), but the results cannot be compared because no such other publication is being reported mentioning the effect of light and dark conditions. The incubation temperature for endolichenic fungi should reflect natural conditions and ranges from 21 to 26 °C.

Sub-culturing of Endolichenic Fungi Plates should be monitored regularly and the rapidly emerging colonies should be isolated and sub-cultured on to different plates as quickly as possible onto media without inhibitors to enhance normal sporulation for better identification. Sub-culturing the primary fungal colony helps to prevent the overlapping of mycelia of two different colonies. The most common media used for sub-culturing the endolichenic fungi is Malt extract agar.

Preservation of Cultures The preservation of fungal cultures is needed to secure the viability and morphological, physiological and genetic integrity of the cultures over time. Preservation methods are of two types, one is short-term and the other is long-term preservation. The selection of the method depends on the concerned species, the available resources and the aim of the study. According to article 8.4 of International Code of Nomenclature for algae, fungi, and plants (=International Code of Botanical Nomenclature) the permanently preserved cultures in metabolically inactive states can serve as type specimens for further studies (Greuter et al. 2000), so it is recommended that the fungal cultures should be preserved by the long-term preservation methods.

Some Points Which Need to Be Kept in Mind While Isolating Endolichenic Fungi

- The fragment size of surface sterilized lichen thallus has nothing to do with the diversity of endolichenic fungi, since till date no such correlation between fungal diversity and thallus size is known. According to the available literature there is no direct or indirect effect of size of fragment, mounted on the media, on the diversity of the endolichenic fungal population acquired from any particular lichen.
- The age is not the case with lichens because an average size thallus of lichen can be more than 50 years old. Till now not a single literature is available mentioning about the age of the thallus from which they have isolated endolichenic fungi.
- Extensive collection of different lichen species from a particular geographical area can recover a great diversity of endolichenic fungi and in itself is a more time and cost effective way for surveying the fungal wealth.
- Extensive collection of a selective species growing on a particular host (e.g. rock, soil, or bark) from ecologically varied sites will also be a good method to obtain diverse fungal populations.
- The diversity of endophytes (strains and species) depends on how much care, time and petri plates are used in a study; a meticulous researcher will laboriously isolate thousands of strains and consequently more species; a lackadaisical researcher will achieve the opposite.
- The diversity of endolichenic fungi isolated from lichens growing in temperate regions will be different from those of tropical regions
- Amendment in the surface sterilization protocols, culture methods, segment size to be used, composition of the culture media will also surely affect the diversity of the endolichenic fungi.

5 Isolation of Secondary Metabolites

The process of obtaining secondary metabolites from fungal strains involves two steps – (1) fermentation and, (2) isolation of secondary metabolites produced during fermentation. Fermentation is prime requisite in the production of secondary metabolites from fungal strains and involves first taking the master culture and preparing a series of sub-master slopes on a nutrient agar such as potato dextrose agar. When these slopes have grown satisfactorily, a seed culture is prepared using a sterile medium in a plugged conical flask, and when this seed culture has grown usually for about 48 h; aliquots are distributed through the production vessels. If the fermentation is to be grown on surface culture, a sterile shallow layer of the medium is inoculated in a flat one-litre Roux bottle. The offset neck of the Roux bottle should be plugged with cotton wool to allow the fermentation to breathe whilst avoiding contamination from spores in the air. Number of bottles inoculated depends on the amount of secondary metabolite one requires. The use of large conical flasks, though being quite bulky, are more convenient, particularly if the fungus is being grown on a solid support such as rice, sterile wood chips, filter paper or glass wool impregnated with the liquid medium. The conical flasks are used for shake culture in which the fermentation is grown on an oscillating table. The shaking movement increases the aeration and dispersion of the mycelium. As the fermentation progresses, some parameters should be kept in mind and need to be regularly monitored, *viz.* sugar level (which can be measured by the optical rotation of an aliquot), the pH, mycelial dry weight, oxygen levels and the availability of nitrogen source, etc. Whilst a surface culture may be grown for a month or even longer, a shake culture may be grown for 10–14 days and a stirred fermentation for an even shorter period of time. Methods have been developed for continuous culture in which fresh sterile medium is added to a growing fermentation whilst spent medium containing the metabolites is removed. Throughout these transfers and fermentation process one has to be very much careful in order to avoid contamination, which can take place by any of the reasons, such as 'flaming' surfaces such as the wire loops and the neck of the flask, which may have come in contact with potential infection; several organisms are being grown in the same laboratory; the collections from the wild are kept in the culture room; the spent fermentation medium and residues need to be treated with a disinfectant or autoclaved and disposed of hygienically as they can still provide nutrients for bacteria and fungi to grow.

After fermentation gets completed the process of isolation of secondary metabolites is performed. There are various ways of harvesting fermentation and isolating the metabolites. But before going into that process one should be aware that there are some metabolites which are excreted into the culture broth whilst others are unable to do so and are retained in the mycelium. Besides this, the metabolites found in the medium/broth could be different in comparison to those found in the mycelium. For example, the sesquiterpenoid metabolite of *Trichothecium roseum*, trichothecin, is found mainly in the culture filtrate whilst the diterpenoid rosenonolactone is found in the mycelium of the same fungus. As a crude generalization, the extra-cellular metabolites isolated from the culture filtrate may be associated with the combative relationship of the organism with its environment, whilst those isolated from the mycelium may have a protective role.

Mycelial metabolites are best isolated by filtering the mycelium, drying it, preferably with gentle heat under vacuum, and then extracting the powdered mycelium in a Soxhlet thimble with a solvent such as chloroform. However, the metabolites from the broth are extracted with a solvent such as diethyl ether, ethyl acetate, butyl acetate or chloroform. There is, sometimes, a loose association between a metabolite and protein in the culture filtrate and the recovery of the metabolite may be improved by acidifying the broth to pH 2. However, care has to be taken not to generate artifacts arising from acid-catalyzed reactions. An alternative method, particularly when large volumes are involved, is to add active charcoal (ca. 12 g/L) to the culture filtrate and then leave the mixture to stand in a cold-room overnight. The charcoal is then filtered off and the metabolites can be eluted from the charcoal with acetone. The metabolites are then separated through the standard natural product techniques by dividing them into acidic, neutral and basic fractions followed by chromatography. Typical yields of the major metabolites obtained from laboratory cultures, as opposed to industrial fermentations, are of the order of 50–100 mg/L of the culture broth. However, in some cases the yield of an interesting metabolite may be as low as 1 mg/L and this could be due to the fact that after repeated sub-culturing, a fungus ceases to produce a secondary metabolite. This is not surprising given the role of some secondary metabolites in facilitating the growth of the fungus in a competitive environment.

References

Girlanda M, Isocrono D, Bianco C et al (1997) Two foliose lichens as microfungal ecological niches. Mycologia 89:531–536

Greuter W, McNeill J, Barrie FR et al (eds and comps) (2000) International code of botanical nomenclature (Saint Louis Code). Regnum Veg 138:1–474

Li WC, Zhou J, Guo SY et al (2007) Endophytic fungi associated with lichens in Baihua mountain of Beijing, China. Fungal Divers 25:69–80

Petrini O, Hake U, Dreyfuss M (1990) An analysis of fungal communities isolated from fruticose lichens. Mycologia 82:444–451

Schulz B, Wanke U, Draeger S et al (1993) Endophytes from herbaceous plants and shrubs-effectiveness of surface sterilization methods. Mycol Res 97:1447–1450

Suryanarayanan TS, Thirunavukkarasu N, Hariharan GN et al (2005) Occurrence of non-obligate microfungi inside lichen thalli. Sydowia 57(1):120–130

Tripathi M, Joshi Y (2015) Endolichenic fungi in Kumaun Himalaya: a case study. In: Upreti DK, Divakar PK, Shukla V et al (eds) Recent advances in lichenology. Springer, New Delhi, pp 111–120

Methods for Identification of Endolichenic Fungi

Abstract Various estimates about the fungal diversity suggest that there is an immense diversity of fungi which is yet to be discovered. To report the correct number of fungal diversity, first one has to isolate and identify them correctly from their host/substrate. To identify the fungi there are different ways available i.e. with the help of morphological features, through biochemical means, anatomical features and the latest one in vogue is the molecular characterization. All the techniques are very helpful to identify the fungi and different researchers use different methods for identification according to the facility available at their host institute or outsource the facility which is needed by them and not available at their institute. In this chapter we have tried to mention all the different methods used for the identification of endolichenic fungi.

Fungi are one of the most diverse kingdoms with an estimated global wealth of 5.1 million number of species, (Hawksworth 1991, 2001; Hawksworth and Rossman 1997; O'Brien et al. 2005) out of which about 99,000 species have been described (Blackwell 2011; Lee et al. 2010) with a fascinating rate of around 1000 species being described each year and rest are in the dark. If the discovered fungal diversity (99,000) is considered to be an iceberg then the number of the fungi grown successfully in pure culture (*ca.* 7000) should be represented by the tip of the iceberg (Hawksworth 1988). Since some fungi are specific to a certain host, hence the isolation of such fungi from different and unusual hosts, such as lichens, with the help of appropriate methodology, can result in culturing many more species ensuing in escalating the size of this tip. Discovering the novel species and growing them in pure culture is incredibly valuable for the mankind as fungi are the largest drug producer in nature. The number of fungal species grown in pure culture with some significant bioactive potential is directly proportional to their probable biotechnological implications. In comparison to the staggering 5.1 million estimated 99,000 discovered fungal species appear to be a very tiny number and one of the several reasons behind this huge gap is the lack of well-trained mycologists and correct identification.

Fungal identification requires significant observation, avoiding this may lead to confusion and wrong identification which further can lead to research workers wast-

M. Tripathi, Y. Joshi, *Endolichenic Fungi: Present and Future Trends*,
https://doi.org/10.1007/978-981-13-7268-1_4

ing scarce resources searching for the same properties in the wrong strains, or establishing research programs based on them. Wrong identification itself doesn't seem so bad but suppose if a strain has been identified earlier as 'A' and now it is being wrongly identified as 'B' then the biotechnologist (non-mycologist) e.g. would evaluate its anticancer potential, which has already been done earlier. If the erroneously identified strain gets deposited in a culture collection then it becomes the permanent identity of that particular strain because culture collections have a heavy workload and also not all the culture collections have the sufficient number of specialist to cross verify the identity of the deposited strain and this leads to the same destination mentioned above.

Correct identification of an unknown isolate generally requires the presence of teleomorphic (sexual state, formerly called *perfect state)* or anamorphic (asexual state, formerly called *imperfect* state) spore stages (Hennebert and Weresub 1977). Manipulation of cultural conditions may stimulate the production of such stages in some strains; however, the absence of teleomorphic spore stages may be due to suboptimal cultural conditions or heterothallism, a pattern of sexuality in which strains are self-sterile. Even though homothallic strains are self-fertile, teleomorphic stages have not been observed in pure culture for many homothallic taxa. The conidial state is the predominant reproductive stage in laboratory culture and, where known, most conidial anamorphs represent asexual stages of ascomycetous teleomorphs.

Various methods have been utilized to investigate diversity of endophytic fungi; however, the conventional cultivation-dependent methods are being quite commonly utilized till date. The cultivation-dependent methods comprise three steps: (1) disinfection of sample surface, (2) isolation of endophyte, and (3) identification of endophyte. Though cultivation-dependent technique is very advantageous as it allows the quick isolation of a great amount of endophytes but has some limitations too: (1) it is painstaking and time taking and inappropriate in comparing huge numbers of samples, (2) identification of sterile isolates to any taxonomic category is also a unique predicament, and (3) the fungi which fail to appear on growth media may be left uncounted from the endophytic population because they might be effortlessly outcompeted by rapidly-growing species on growth media.

The morphotaxonomic identification of endophytic fungi isolated by the traditional cultivation-dependent process can be possible if they sporulate on the media. Regardless of the attempts to promote sporulation (Guo et al. 1998, 2000a, b; Taylor et al. 1999), the number of non-sporulating isolates varies from 4.5 to 54% of the total isolates (Petrini et al. 1982; Espinosa-Garcia and Langenheim 1990; Johnson and Whitney 1992; Fisher et al. 1993; Guo et al. 2000a, b, 2008; Photita et al. 2001; Cannon and Simmons 2002; Kumaresan and Suryanarayanan 2002; Wang and Guo 2007; Sun et al. 2008). Since the traditional system of classification of fungi is solely based on fruiting bodies hence, it does not have any system for the taxonomic naming of sterile isolates. By keeping in mind the enormous diversity of sterile isolates, now they are being classified as "morphotype" on the basis of their phenotypical characters (Taylor et al. 1999; Guo et al. 2000a, b, 2003; Arnold et al. 2001;

Ward et al. 2005). The classification of taxa into various morphotypes, though, does not exhibit species phylogeny, since morphotypes are not genuine taxonomic units (Guo et al. 2000a, b, 2003; Lacap et al. 2003; Ward et al. 2005).

Therefore, to resolve the prospective technical prejudice, cultivation-independent methods, e.g. molecular approaches, such as amplified fragment length polymorphism (AFLP), denaturing gradient gel electrophoresis (DGGE), random amplified polymorphic DNA (RAPD), restriction fragment length polymorphism (RFLP), single-stranded conformation polymorphism (SSCP), simple sequence repeat (SSR), temperature gradient gel electrophoresis (TGGE) and terminal-RFLP (T-RFLP) have recently been adopted and applied to identify and understand the diversity of these endophytic mycelia sterilia and population structure and community dynamics of endophytic fungi (McCutcheon and Carroll 1993; Groppe et al. 1995; Groppe and Boller 1997, De Jong et al. 2003, Grünig et al. 2001, 2002, 2004, 2006, 2007, 2008; Duong et al. 2006; Ranjard et al. 2000; Kuhls et al. 1996; Gernandt et al. 1997).

Estimation of phylogenetic position of any sterile isolate can be achieved with the help of Nucleic acid sequencing. Constructing partial phylogenies of acso- and basidio-mycetous fungi is made possible through the use of sequence analysis of amplified DNA copies of various regions of ribosomal RNA (rRNA) genes isolated from targeted isolates (Bruns et al. 1991; Berbee and Taylor 1992a, b; Carbone and Kohn 1993; Zambino and Szabo 1993; Swann and Taylor 1993, 1995a, b, c; Monreal et al. 1999). The systematic analysis of similarity between the nucleotide sequences, an unknown sterile isolate can be provided a taxonomic status (order, family, and even sometimes genus). By following this method of characterization the endolichenic fungal diversity can be estimated without being bothered about the absence of sporulation in the isolates. When the isolate has no identity it is hard to tell that what kinds of growth conditions are favorable for it because one doesn't know which group it belongs but the nucleotide sequence analysis doesn't need any fruiting body to identify the fungi. So through this technique the investigator can know the proximate taxonomical status of the taxon and then select the conditions that will regulate growth and sporulation in that particular taxon. The use of molecular analyses to establish connections between anamorphs and teleomorphs (LoBuglio et al. 1993; Rehner and Samuels 1994), as well as phylogenetic relationships of autonomous anamorphs and closely related teleomorph genera, has now become a routine practice. Since new techniques used to develop rapidly, the authors hereby are not recommending any specific methods for identification of endophytes based on amplified sequences, but in general, are providing the means by which they can be identified. Besides this, endophytes as such comprise a large and diverse group of fungi, so no identification methods will apply to endophytes in general. The collective effort of morphological, biochemical and molecular tools could lead to the authenticity of strains, all of which are described here in detail. Hence, to avoid such incidents, authentication and establishment of purity of strains should be done with the collective effort of morphological and molecular tools, which are described here.

1 Morphotaxonomic Approach

Taxonomy of any fungal species on the basis of morphological appearance is the primary, significant, informative, cost effective and easy to perform approach. But as mentioned earlier, this method is applicable only in those fungi which bear spores or conidia. The identification of endolichenic fungi starts with the examination of colonial morphology of fungi *viz.* colony color, size, shape, texture, edges. The second step involves making a temporary mount of the fungi present in the culture plate which could be done with ease and comfort by using scotch tape technique (Harris 2000). In this incredibly simple method the tape is cut in an appropriate length and tapped on to pure culture of fungi and an impression is made and then this tape with impressions of fungi is placed on a glass slide having a drop of lacto-phenol cotton blue. Then this preparation is observed under compound microscope. The observer should notice septation in mycelia, color of mycelia, attachment of spore with hyphae, spore size, color, shape, septation in spore and notes should be made. Spore sizes can vary considerably, even within single strains, so it is always advisable to examine at least 20 mature spores to obtain an accurate measurement. In case of conidial fungi (Deuteromycotina), the gross appearance of colonies developed on medium is of considerable importance in identification. Prior to all these characters, the culture medium, temperature, pH, and period of incubation should also be noted down as they can affect the colony morphology and sporulation. Hence, it is quite important to employ standard procedures and media recommended for particular fungal genera. The isolates which do not sporulate in culture should be treated as mycelia sterilia. The fungi could be identified using relevant keys and taxonomic notes.

2 Biochemical Approach

The biochemical characterization of fungi which can be easily performed by standardized thin layer chromatographic (TLC) procedures (Frisvad 1981; Paterson 1986) is not only of interest to biotechnologists, but to those who are concerned with the differentiation of pathogenic from non pathogenic, or toxic from non-toxic strains of the same fungus species. It also helps in understanding the host preferences and substrate utilization and could be related to the enzymes which are produced in the culture. The secondary metabolites of some genera have been studied quite thoroughly and the presence or absence of particular compounds leads to the identification of species (Table 1).

Table 1 Examples of physiological and biochemical activities used in the characterization of fungus strains (Hawksworth 1988)

Growth in presence of inhibitors (e.g. formalin, preservatives, copper sulphate, phenol, sodium azide, malachite green)
Growth on low-water activity media, at different temperatures and at different pH
Utilization of carbon sources (e.g. lactose, glucose, soluble starch, sucrose, mannitol, oxalate)
Utilization of nitrogen sources (e.g. nitrate, nitrite, ammonium, creatine, glycine)
Enzyme activities; hydrolysis (e.g. aesculin, starch, tween 80, cellulose, lignin, RNA, casein, gelatin); reduction (of e.g. tetrazolium, tellurite);
APIZYM tests (e.g. aryeomidases, chymotrypsinase, galactosidases); isoenzymes (e.g. pectinases)
Secondary metabolite profiles; production of mycotoxins and other compounds, separated and visualized by thin-layer chromatography

3 Molecular Approach

DNA Extraction, Amplification, and Sequencing The isolates from each lichen species should at first be arranged into morphospecies according to their phenotypic characteristics of colony (e.g. color and texture) and then further precede to molecular identification. Because the nuclear internal transcribed spacers (ITS) region (ITS1-5.8 S-ITS2) of the nrRNA gene is the default marker for study of fungi at the species level, sequences of ITS region can be used for fungal identification. The genomic DNA can be extracted with CTAB method as explained by Cubero et al. (1999) and the ITS region should be amplified with the primers ITS1F and ITS4 as described by White et al. (1990). Additionally, partial region of 28S rRNA gene can be amplified with the primers LR3R and LR7 (http://www.biology.duke.edu/fungi/mycolab/primers.htm). PCR amplification can be performed as follows: 95 °C for 3 min, followed by 37 cycles of 94 °C for 30 s, 52 °C for 30 s, and 72 °C for 30 s; and a final extension at 72 °C for 10 min, and finally the PCR products need to be purified and sequenced from any reliable source. The sequence data obtained should be deposited in GenBank under a certain accession number.

Molecular Identification and Phylogeny After acquiring the nucleotide sequence data, the isolates need to be identified on the basis of similarity between sequences and the phylogenetic position in the phylogenetic tree. To determine similarity between sequences, ITS sequences of fungal isolates should be submitted to the BLASTN program (http://blast.ncbi.nlm.nih.gov/Blast.cgi) to establish the close matches in GenBank database. The following criterion is generally used to interpret the fungal sequences of the GenBank database: for ITS sequence identities ≥99%, the fungal species were accepted; for ITS sequence identities of 95–99%, only the genus was accepted; for ITS sequence identities <95%, OTUs were labeled at the order, family or phylum name or as 'fungus'. Phylogenetic analysis of fungal species could be performed using MEGA software v. 6.0 and neighbor-joining phylogenetic trees can be constructed using the ITS sequences and partial 28S rRNA gene sequences. The maximum composite likelihood method can be employed to assess the evolutionary distances with bootstrap values calculated from 1000 replicate runs.

Pyrosequencing Recently this technique has been applied by Zhang et al. (2015) for isolating endolichenic fungi and under this technique the fungal internal transcribed spacer (ITS, ITS1–5.8S-ITS2) of nuclear ribosomal DNA sequences was amplified by using ITS1F (55′ CTTGGTCATTTAGAGGAAGTAA -3′) and ITS4 (55′- TCCTCCGCTTATTGATATGC -3′). The PCR amplification is performed using the Amplicon Fusion Primers as 55′ A-x-ITS1F-3′ and 55′-B-ITS4–3′, where A and B represents the pyrosequencing adaptors and x represent 8 bp-tag for sample identification. The reaction mixture contains the template DNA (10 ng of Template DNA), 5 × buffer (50 M Tris-HCl, pH 8.3–8.8), 2.5 nM dNTP, Fastpfu Polymerase, each primer and ultra pure sterilized water. The PCR amplification consisted of an initial denaturation at 95 °C for 2 min, 30 cycles of denaturation at 95 °C for 30 s, annealing at 55 °C for 30 s, and extension at 72 °C for 30 s, with a final extension at 72 °C for 5 min. After amplification, it was purified using the AxyPrep DNA Gel Extraction Kit (Axygen Biosciences, Inc., US) and quantified using QuantiFluor-ST (Promega Corporation, US). Finally the equimolar mixtures of multiple amplicons were used for pyrosequencing on a Roche 454 GS FLX + Titanium platform (Roche 454 Life Sciences, US). The raw sequence reads were deposited in the NCBI sequencing read archive under a certain accession number. The raw sequence data generated during the above procedure was processed using QIIME 1.8.0 software43. Briefly, the sequence libraries were split and denoised to avoid diversity overestimation caused by sequencing errors, including sequences with average quality score < 20 over a 50 bp sliding window, sequences shorter than 200 bp, sequences with homopolymers longer than six nucleotides, and sequences containing ambiguous base calls or incorrect primer sequences. Operational Taxonomic Units (OTUs) were then clustered with a 97% similarity cutoff using UPARSE44 and chimeric sequences were identified and removed using UCHIME45. These OTUs can be used as a basis for calculating alpha-diversity and beta-diversity metrics.

4 Conclusion

Nowadays PCR-based molecular techniques are quite conventional and prevalent to identify the endolichenic fungi, especially for sterile isolates, and for the characterization of the viable but non-culturable fungi by directly amplified rDNA fragments from plant tissues. But to make these markers work in case of lichens, regarding the identification of non-culturable fungi, they need to be made more advanced. The traditional PCR engaged in the recognition and characterization of endophytic fungi cannot be used to quantify biomass of endophytic fungi. Nevertheless, real-time PCR which can sense little amount of DNA in environmental samples can be productively used to establish the population density of few species of fungi (Bates et al. 2001; Atkins et al. 2003, 2004; Zhang et al. 2006). DNA barcoding technique which uses a quick, efficient, standardized gene region to determine the species (Hebert et al. 2003; Blaxter 2003; Savolainen et al. 2005; Seifert et al. 2007; Craig et al. 2008) has being widely employed in the animal kingdom by using a 648-bp region of the CO1 gene (Smith et al. 2005; Ward et al. 2005; Hajibabaei et al. 2006). Its use in fungal identification has been started recently and was engaged in the characterization of *Penicillium* species through the use of CO1 gene (Seifert et al. 2007) and ectomycorrhizal fungi in a Tasmanian wet sclerophyll forest by using ITS regions (Tedersoo et al. 2008). This technique needs to be widely implemented in the field of endophytic fungi. Hence, with the use of DNA barcoding, DNA fingerprinting, DNA sequencing and real-time PCR, rapid characterization and quantification of endolichenic fungal diversity will be possible in near future.

References

Arnold AE, Maynard Z, Gilbert GS (2001) Fungal endophytes in dicotyledonous neotropical trees: patterns of abundance and diversity. Mycol Res 105:1502–1507

Atkins SD, Clark IM, Sosnowska D et al (2003) Detection and quantification of *Plectosphaerella cucumerina*, a potential biological control agent of potato cyst nematodes, by using conventional PCR, real-time PCR, selective media and baiting. Appl Environ Microbiol 69:4788–4793

Atkins SD, Clark IM, Pande S et al (2004) The use of real-time PCR and species specific primers for the identification and monitoring of *Paecilomyces lilacinus*. FEMS Microbiol Ecol 51:257–264

Bates JA, Taylor EJA, Kenyon DM et al (2001) The application of real-time PCR to the identification, detection and quantification of *Pyrenophora* species in barley seed. Mol Plant Pathol 2:49–57

Berbee ML, Taylor JW (1992a) Convergence in ascospore discharge mechanism among pyrenomycetous fungi based on 18S ribosomal RNA gene sequence. Mol Phylogenet Evol 1:59–71

Berbee ML, Taylor JW (1992b) Two ascomycete classes based on fruiting-body characters and ribosomal DNA sequence. Mol Biol Evol 9:278–284

Blackwell M (2011) The fungi: 1, 2, 3 … 5.1 million species? Am J Bot 98(3):426–438

Blaxter M (2003) Counting angels with DNA. Nature 421:122–124

Bruns TD, White TJ, Taylor JW (1991) Fungal molecular systematics. Annu Rev Ecol Syst 22:525–564

Cannon PF, Simmons CM (2002) Diversity and host preference of leaf endophytic fungi in the Iwokrama Forest Reserve, Guyana. Mycologia 94:210–220

Carbone I, Kohn LM (1993) Ribosomal DNA sequence divergence within internal transcribed spacer 1 of the Sclerotiniaceae. Mycologia 85:415–427

Craig DW, Pearson JV, Szelinger S et al (2008) Identification of genetic variants using bar-coded multiplexed sequencing. Nat Methods 5:887–893

Cubero OF, Crespo A, Fathi J et al (1999) DNA extraction and PCR amplifcation method suitable for fresh, herbarium-stored, lichenized and other fungi. Plant Syst Evol 216:243–249

De Jong EZ, Guthridge KM, Spangenberg GC et al (2003) Development and characterization of EST-derived simple sequence repeat (SSR) markers for pasture grass endophytes. Genome 46:277–290

Duong LM, Jeewon R, Lumyong S et al (2006) DGGE coupled with ribosomal DNA gene phylogenies reveal uncharacterized fungal phylotypes. Fungal Divers 23:121–138

Espinosa-Garcia FJ, Langenheim JH (1990) The leaf fungal endophytic community of a coastal redwood population diversity and spatial patterns. New Phytol 116:89–97
Fisher PJ, Petrini O, Sutton BC (1993) A comparative study of fungal endophytes in leaves, xylem and bark of *Eucalyptus nitens* in Australia and England. Sydowia 45:338–345
Frisvad JC (1981) Physiological criteria and mycotoxin production as aids in identification of common asymmetric penicillia. Appl Environ Microbiol 41:568–579
Gernandt DS, Camacho FJ, Stone JK (1997) *Meria laricis*, an anamorph of *Rhabdocline*. Mycologia 89:735–744
Groppe K, Boller T (1997) PCR assay based on a microsatellite-containing locus for detection and quantification of *Epichloë* endophytes in grass tissue. Appl Environ Microbiol 63:1543–1550
Groppe K, Sanders I, Wiemken A et al (1995) A microsatellite marker for studying the ecology and diversity of fungal endophytes (*Epichloë* spp.) in grasses. Appl Environ Microbiol 61:3943–3949
Grünig CR, Sieber TN, Holdenrieder O (2001) Characterisation of dark septate endophytic fungi (DSE) using inter-simple-sequence-repeat-anchored polymerase chain reaction (ISSR-PCR) amplification. Mycol Res 105:24–32
Grünig CR, Sieber TN, Rogers SO et al (2002) Spatial distribution of dark septate endophytes in a confined forest plot. Mycol Res 106:832–840
Grünig CR, McDonald BA, Sieber TN et al (2004) Evidence for subdivision of the root-endophyte *Phialocephala fortinii* into cryptic species and recombination within species. Fungal Genet Biol 41:676–687
Grünig CR, Duò A, Sieber TN (2006) Population genetic analysis of *Phialocephala fortinii* s.l. and *Acephala applanata* in two undisturbed forests in Switzerland and evidence for new cryptic species. Fungal Genet Biol 43:410–421
Grünig CR, Brunner PC, Duò A et al (2007) Suitability of methods for species recognition in the *Phialocephala fortinii-Acephala applanata* species complex using DNA analysis. Fungal Genet Biol 44:773–788
Grünig CR, Duò A, Sieber TN, Holdenrieder O (2008) Assignment of species rank to six reproductively isolated cryptic species of the *Phialocephala fortinii* s.l.-*Acephala applanata* species complex. Mycologia 100:47–67
Guo LD, Hyde KD, Liew ECY (1998) A method to promote sporulation in palm endophytic fungi. Fungal Divers 1:109–113
Guo LD, Hyde KD, Liew ECY (2000a) Identification of endophytic fungi from *Livistona chinensis* (Palmae) using morphological and molecular techniques. New Phytol 147:617–630
Guo LD, Hyde KD, Liew CY (2000b) Identification of endophytic fungi from *Livistona chinensis* based on morphology and rDNA sequences. New Phytol 147:617–630
Guo LD, Huang GR, Wang Y et al (2003) Molecular identification of white morphotype strains of endophytic fungi from *Pinus tabulaeformis*. Mycol Res 107:680–688
Guo LD, Huang GR, Wang Y (2008) Seasonal and tissue age influences on endophytic fungi of *Pinus tabulaeformis* (Pinaceae) in Dongling Mountain, Beijing. J Integr Plant Biol 50:997–1003
Hajibabaei M, Janzen DH, Burns JM et al (2006) DNA barcodes distinguish species of tropical Lepidoptera. PNAS 103:968–971
Harris JL (2000) Safe, low-distortion tape touch method for fungal slide mounts. J Clin Microbiol 38:4683–4684
Hawksworth DL (1988) Identification. In: Hawksworth DL, Kirsop BE (eds) Living resources for biotechnology: filamentous fungi. Cambridge University Press, Cambridge, pp 100–114
Hawksworth DL (1991) The fungal dimension of biodiversity: magnitude, significance, and conservation. Mycol Res 95:641–655
Hawksworth DL (2001) The magnitude of fungal diversity: the 1.5 million species estimate revisited. Mycol Res 105:1422–1432
Hawksworth DL, Rossman AY (1997) Where are all the undescribed fungi? Phytopathology 87:888–891
Hebert PDN, Ratnasingham S, deWaard JR (2003) Barcoding animal life: cytochrome c oxidase subunit 1 divergences among closely related species. Proc R Soc B 270(Suppl):96–99

Hennebert GL, Weresub LK (1977) Terms for states and forms of fungi, their names and types. Mycotaxon 6:207–211
Johnson JA, Whitney NJ (1992) Isolation of fungal endophytes from black spruce (*Picea mariana*) dormant buds and needles from New Brunswick. Can J Bot 70:1754–1757
Kirk PM, Cannon PF, Minter DW et al (2008) Dictionary of the fungi, 10th edn. CABI, Wallingford
Kuhls K, Lieckfeldt E, Samuels GJ et al (1996) Molecular evidence that the asexual industrial fungus *Trichoderma reesei* is a clonal derivative of the ascomycete *Hypocrea jecorina*. Proc Natl Acad Sci USA 93:7755–7760
Kumaresan V, Suryanarayanan TS (2002) Endophytic assemblages in young, mature and senescent leaves of *Rhizophora apiculata*: evidence for the role of endophytes in mangrove litter degradation. Fungal Divers 9:91–91
Lacap DC, Hyde KD, Liew ECY (2003) An evaluation of the fungal 'morphotype' concept based on ribosomal DNA sequences. Fungal Divers 12:53–66
Lee SC, Corradi N, Doan S et al (2010) Evolution of the sex-related locus and genomic features shared in microsporidia and Fungi. PLoS One 5:e10539. https://doi.org/10.1371/journal.pone.0010539
LoBuglio KF, Pitt JI, Taylor JW (1993) Phylogenetic analysis of two ribosomal DNA regions indicates multiple independent losses of a sexual *Talaromyces* state among asexual *Penicillium* species in subgenus *Biverticillium*. Mycologia 85:592–604
McCutcheon TL, Carroll GC (1993) Genotypic diversity in populations of a fungal endophyte from Douglas fir. Mycologia 85:180–186
Monreal M, Berch SM, Berbee M (1999) Molecular diversity of ericoid mycorrhizal fungi. Can J Bot 77:1580–1594
O' Brien BL, Parrent JL, Jackson JA et al (2005) Fungal community analysis by large-scale sequencing of environmental samples. Appl Environ Microbiol 71:5544–5550
Paterson RRM (1986) Standardized one- and two-dimensional thin layer chromatographic methods for the identification of secondary metabolites in *Penicillium* and other fungi. J Chromatogr 368:249–264
Petrini O, Stone J, Carroll FE (1982) Endophytic fungi in evergreen shrubs in western Oregon: a preliminary study. Can J Bot 60:789–796
Photita W, Lumyong S, Lumyong P et al (2001) Endophytic fungi of wild banana (*Musa acuminata*) at doi Suthep Pui National Park, Thailand. Mycol Res 105:1508–1513
Ranjard L, Poly F, Nazaret S (2000) Monitoring complex bacterial communities using culture-independent molecular techniques: application to soil environment. Res Microbiol 151:167–177
Rehner SA, Samuels GJ (1994) Taxonomy and phylogeny of *Gliocladium* analysed from nuclear large subunit ribosomal DNA sequences. Mycol Res 98:625–634
Savolainen V, Cowan RS, Vogler AP et al (2005) Towards writing the encyclopaedia of life: an introduction to DNA barcoding. Philos Trans R Soc B 360:1805–1811
Seifert KA, Samson RA, de Waard JR et al (2007) Prospects for fungus identification using CO1 DNA barcodes, with *Penicillium* as a test case. PNAS 104:3901–3906
Smith MA, Fisher BL, Hebert PDN (2005) DNA barcoding for effective biodiversity assessment of a hyperdiverse arthropod group: the ants of Madagascar. Philos Trans R Soc B 360:1825–1834
Sun JQ, Guo LD, Zang W et al (2008) Diversity and ecological distribution ofendophytic fungi associated with medicinal plants. Sci China Ser C Life Sci 51:751–759
Swann EC, Taylor JW (1993) Higher taxa of Basidiomycetes: an 18S rRNA gene perspective. Mycologia 85:923–936
Swann EC, Taylor JW (1995a) Phylogenetic perspectives on basidiomycete systematics: evidence from the 18S rRNA gene. Can J Bot 73(Suppl 1):S862–S868
Swann EC, Taylor JW (1995b) Toward a phylogenetic systematics of the Basidiomycota: integrating yeast and filamentous basidiomycetes using 18S rRNA gene sequences. Stud Mycol 38:147–161
Swann EC, Taylor JW (1995c) Phylogenetic diversity of yeast-producing basidiomycetes. Mycol Res 99:1205–1210

Taylor JE, Hyde KD, Jones EBG (1999) Endophytic fungi associated with the temperate palm, *Trachycarpus fortunei*, within and outside its natural geographic range. New Phytol 142:335–346
Tedersoo L, Jairus T, Horton BM et al (2008) Strong host preference of ectomycorrhizal fungi in a Tasmanian wet sclerophyll forest as revealed by DNA barcoding and taxon-specific primers. New Phytol 180:479–490
Wang Y, Guo LD (2007) A comparative study of endophytic fungi in needles, bark, and xylem of *Pinus tabulaeformis*. Can J Bot 85:911–917
Ward RD, Zemlak TS, Innes BH et al (2005) DNA barcoding Australia's fish species. Philos Trans R Soc B 360:1847–1857
White TJ, Bruns T, Lee S et al (1990) Amplification and direct sequencing of fungal ribosomal RNA genes for phylogenetics. In: Innis MA, Gelfand DH, Sninsky JJ et al (eds) PCR protocols, a guide to methods and applications. Academic, New York, pp 315–322
Zambino PJ, Szabo LJ (1993) Phylogenetic relationships of selected cereal and grass rusts based on rDNA sequence analysis. Mycologia 85:401–414
Zhang N, Castlebury LA, Miller AN et al (2006) An overview of the systematic of sordariomycetes based on four-gene phylogeny. Mycologia 98:1076–1087
Zhang T, Wei XL, Zhang YQ et al (2015) Diversity and distribution of lichen associated fungi in the Ny-Ålesund region (Svalbard, high Arctic) as revealed by 454 pyrosequencing. Sci Rep 5:14850

Endolichenic Fungi: A Promising Source for Novel Bioactive Compounds

Abstract A large proportion of the drugs available in market are either of plant or microbial origin. Microbes produce various secondary metabolites and these metabolites turned out to be biologically active. Endophytic fungi are considered to be more creative in case of producing novel secondary metabolites. Lichens also harbor endophytes i.e. endolichenic fungi, which produces a large number of metabolites (alkaloids, cyclic peptides, polyketides, steroids and terpeniods) including a great proportion of novel metabolites displaying a wide spectrum of bio-activities. A great diversity of lichens is known across the world of which only a lichens have been screened for their endolichenic fungal diversity and secondary metabolites. So a huge proportion of work is left to be completed regarding the secondary metabolites of endolichenic fungi. This chapter entails all the possible metabolites of endolichenic fungi known so far and their bioactive potential.

Natural products are an imperative supply of novel pharmaceutical products (Dreyfuss and Chapela 1994; Proudfoot 2002) and to our surprise in excess of 60% anticancer and 70% antimicrobial drugs at present in medical practice are either natural products or natural product derivatives (McAlpine et al. 2005). Amongst all the known organisms, microorganisms are the vital source of bioactive compounds with colossal prospective to unearth novel molecules for drug discovery, industrial use and agricultural applications (Demain 1999; Keller et al. 2005; Strobel 2006; Porras-Alfaro and Bayman 2011). The metabolites derived from microbes have proven to be more amenable to biosynthetic studies in comparison to natural products obtained from higher plants partly because no seasonal problems have been reported in the production of metabolites and also the incorporation of labeled precursors is quite high (Hanson 2008). Besides this, the microbes are extremely diverse and least investigated. Researches focused on assessment of microbial populations have suggested that merely 1% of bacteria and 5% of fungi have been documented and the others are still anonymous for the benefit of mankind (Heywood 1995; Staley et al. 1997). With regards to that 30% of the regularly used drugs are obtained from fungi (Gloer 1997), and only 5% of the global fungal diversity have been documented (Hawksworth 1991, 2001), fungi propose a massive possibility

M. Tripathi, Y. Joshi, *Endolichenic Fungi: Present and Future Trends*,
https://doi.org/10.1007/978-981-13-7268-1_5

for the discovery of novel pharmaceuticals. In optimizing this pursuit, it is appropriate to bear in mind that the metabolites synthesized by a fungus may be similar to its host taxon and ecological niche, e.g. the mycotoxins of plant pathogens (Dreyfuss and Chapela 1994; Gloer 1997); and metabolic communications may enliven the production of metabolites. Thus, in this context endophytic fungi are one such source for smart screening and can serve as ultimate, readily renewable, and inexhaustible source of novel natural products exhibiting extensive gamut of biological activities.

Throughout the preceding decade, it was perceived that a fair proportion of microbial diversity having novel and bioactive secondary metabolite production exists inside plants (Strobel 2006; Porras-Alfaro and Bayman 2011). Endophytes produce more bioactive secondary metabolites in comparison to the same fungal strain living freely in the environment may be because of their precise purpose in nature and activating a variety of metabolic pathways to endure the host environment (Strobel and Daisy 2003; Strobel 2006; Hassan et al. 2012). Endophytes residing in the plant host involve incessant metabolic interface between fungus and host. In case of secondary metabolite isolation fungal plant pathogens and fungal soil isolates are more worked out in comparison to the endophytic fungi (Tan and Zou 2001). Nearly 80% of endophytic fungi produce secondary metabolites possessing bioactive molecules (Schulz et al. 2002). Not far off from 51% of bioactive secondary metabolites isolated from endophytic fungi were in anonymity in the past (Firakova et al. 2007). These studies draw attention to the substantial capability of endophytic fungi as "factories" of biologically active secondary metabolites.

Majority of the biologically active secondary metabolites are produced by these ostensible 'creative fungi' which comprise species of *Acremonium*, *Aspergillus*, *Fusarium* and *Penicillium*, but a smaller amount of investigations have been performed focusing the novel secondary metabolite production by fungal endophytes. The first break-through regarding endophytic fungi as a source of bioactive compounds came in the year 1993, when researchers in Montana have isolated and identified a fungal species (*Taxomyces andreanae*) that is capable of producing Taxol, a drug currently acquired from the bark of a yew tree (*Taxus* sp.) of limited supply that has shown promise in the treatment of ovarian cancer (Strobel et al. 1996). The discovery of Taxol production by endophytes of Yew tree created the benchmark for carrying out studies on endophytic fungi from various medicinal plants across the world (Tan and Zou 2001; Strobel and Daisy 2003). All these studies are based on the fact that endophytes have evolved with their hosts over a long evolutionary period (Taylor and Taylor 2000) and that they may have changed genetic information with the plants and vice versa (Stierle et al. 1993). Endophytes are well recognized for producing the secondary metabolites which are alike in structure and able to perform similar functions as of the metabolites produced by the host plant (Strobel 2002) and also found producing plant growth regulators (MacMillan 2002). Schulz et al. (2002) stated that fungal endophytes produce about 51% of novel bioactive compounds which proves that screening is not a random walk through a forest; one has to look into the unusual places for the exploration of novel bioactive compounds.

Why endophytes are so much alluring? The answer to this question lies in the traits of endophytes *viz.* diverse, fast growing, creative, and producer of large amount of novel compounds. These characteristic features of endophytes may be accountable for the high diversity of novel bioactive compounds discovered so far. Endolichenic fungi also not lag far behind and this can be justified by the growing body of associated literature on both endolichenic fungal diversity and associated secondary metabolites. Out of 20,000–28,000 lichen species estimated worldwide (Feuerer and Hawksworth 2007; Boonpragob et al. 2012) only 59 lichen species have been screened for their endolichenic fungal diversity till date, which has resulted in the isolation of 104 endolichenic fungi. Not only this, but a total of 351 secondary metabolites have been harnessed from 40 endolichenic fungi isolated from 35 lichen species globally.

During the last decade many strains of endolichenic fungi have been substantially reported producing potential bioactive natural products and novel structures including alkaloids, cyclic peptides, polyketides, steroids and terpeniods exhibiting a variety of pharmacological properties, including antioxidant, anticancer, antifungal, antibacterial, antiviral and anti-Alzheimer's disease properties (Table 1, Fig. 1). The first study performed on secondary metabolites isolated from endolichenic fungi was carried out a decade ago (Paranagama et al. 2007) and time since then over 40 endolichenic fungi have been cultured and subjected to detailed investigations for their secondary metabolites and revealed the occurrence of 351 metabolites out of which 196 were new compounds including polyketides, terpenoids, steroids, alkaloids, and cyclic peptides. As was expected, like endophytic fungi, these are also biologically active (*viz.* $A\beta_{42}$ aggregation inhibitor, cytotoxic against various tumor cell lines, antimicrobial and radical scavenger) (Table 1). Since then this branch of lichenology has grown very well in a steady pace throughout the globe.

Corynesporol (**21**) and some new herbarin derivatives (**22,25,26**) were isolated from *Corynespora* sp. found growing in *Usnea cavernosa* and assessed against PC-3 M and MDA-MB-231 cancer cell lines; none of the isolated metabolites possessed significant inhibitory activity (Paranagama et al. 2007; Wijeratne et al. 2010).

Six new derivatives were isolated from *Pestalotiopsis* sp. inhabiting *Clavaroids* sp. and their antimicrobial efficacy was evaluated against Gram-positive *Streptococcus aureus*. One of the six derivatives (**110**) revealed to be positively active (Ding et al. 2009).

Neurospora terricola, isolated from *Everniastrum cirrhatum* yielded six unique polyketides (**15–20**) having cytotoxic activity against HeLa and MCF-7 tumor cell lines (Zhang et al. 2009).

Five novel compounds [Conioxanthone A (**51**), Conioxepinol A-D (**61–64**)] extracted from cultures of *Coniochaeta* sp. isolated from *Xanthoria mandschurica* were evaluated against HeLa, A-549, MDA-MB-231 tumor cell lines. All four Conioxepinol derivatives (**61–64**) showed cytotoxic activity (Wang et al. 2010a, b).

Wu et al. (2011) revealed the presence of two novel cyclic peptides in *Xylaria* sp. isolated from *Leptogium saturninum*. One of them cyclo(N-methyl-L-Phe-L-Val-D-Ile-L-Leu-L-Pro) (**175**) showed antimicrobial activity against *Candida albicans*.

Table 1 Secondary metabolites isolated from endolichenic fungi and their bioactive properties

Host	Endolichenic fungi	Secondary metabolites	Bioactive properties	References
Hypotrachyna (= *Everniastrum* sp.)	*Alternaria* (= *Ulocladium* sp.)	5′-Methoxy-6-methyl-biphenyl-3,4,3′-triol (**1**)	Radical scavenging activity on DPPH	He et al. (2012), Wang et al. (2012)
Parmelinella wallichiana	*Nigrospora* sp.			
Usnea aciculifera	*Alternaria* sp.			
Cetrelia braunsiana	*Phialophora* sp.			
Hypotrachyna (= *Everniastrum* sp.)	*Ulocladium* sp.	Altenusin (**2**)	Antimicrobial against *Aspergillus fumigatus* and *Bacillus subtilis*	Wang et al. (2012), Ye et al. (2013)
Cladonia ochrochlora	*Phialophora* sp.		Radical scavenging activity on DPPH	
Hypotrachyna (= Everniastrum sp.)	*Alternaria* (= *Ulocladium* sp.)	Alterlactone (**3**)	Antimicrobial against *B. subtilis*	Wang et al. (2012), Huang et al. (2012)
			Radical scavenging activity on DPPH	
Peltigera elisabethae	*Aspergillus* sp.	2-Isopentenyldiorcinol (**4**)[a]		
		Diorcinol (**5**)	Aβ_{42} aggregation inhibitory activity	Huang et al. (2012), Zhao et al. (2014)
		Violaceol-1 (**6**)		Zhao et al. (2014)
		Violaceol-II (**7**)		
		Diorcinol B (**8**)		
		9-Acetyldiorcinol B (**9**)[a]		
Cladonia gracilis	*Scopulariopsis* sp.	1,8-Dimethoxynaphthlen-2-ol (**10**)[a]		Yang et al. (2012)
		1-(40-Hydroxyl-30,50-dimethoxy-phenyl)-1,8-dimethoxynaphthalen-2(1H)-one (**11**)[a]		
Ramalina calicaris	*Chaetomium elatum*	Globosumoside A (**12**)[a]		Chen et al. (2012)
		Globosumoside B (**13**)[a]		

Cetraria islandica	*Myxotrichum* sp.	Myxotrichin D **(14)**	Cytotoxic against K562 cell line	Yuan et al. (2013)
Hypotrachyna cirrhata (= Everniastrum cirrhatum)	*Neurospora terricola*	Terricollene A **(15)**[a]	Cytotoxic against HeLa and MCF-7 cell line	Zhang et al. (2009)
		Terricollene B **(16)**[a]		
		Terricollene C **(17)**[a]	Cytotoxic against HeLa and MCF-7 cell line	
		Terricolyne **(18)**[a]		
		1-*O*-Methylterricolyne **(19)**[a]	Cytotoxic against HeLa and MCF-7 cell line	
		1-*O*-Acetylterricolyne **(20)**[a]		
Usnea cavernosa	*Corynespora* sp.	Corynesporol **(21)**[a]		Paranagama et al. (2007)
		1-Hydroxydehydroherbarin **(22)**[a]		
		Herbarin **(23)**		Paranagama et al. (2007), Wijeratne et al. (2010)
		9-O-Methylscytalol A **(24)**[a]		Wijeratne et al. (2010)
		7-Desmethylherbarin **(25)**[a]		
		8-Hydroxyherbarin **(26)**[a]		
		Scytalol A **(27)**		
		8-O-Methylfusarubin **(28)**		
		Scorpinone **(29)**		
		8-O-Methylbostrycoidin **(30)**		
		1-Methoxydehydroherbarin **(31)**		
Cladina grisea	*Eurotium* sp.	Catenarin **(32)**		Chen et al. (2014)
		Rubrocristin **(33)**		
Ramalina calicaris	*Preussia africana*	Preussochromone B **(34)**[a]		Zhang et al. (2012)
		Preussochromone C **(35)**[a]	Cytotoxic against A-549 and HCT-116 cell line	
		Preussochromone D **(36)**[a]		
		Preussochromone E **(37)**[a]		
		Preussochromone F **(38)**[a]		

(continued)

Table 1 (continued)

Host	Endolichenic fungi	Secondary metabolites	Bioactive properties	References
Hypotrachyna (= Everniastrum sp.)	*Alternaria* (= *Ulocladium* sp.)	Rubralactone **(39)**	Antimicrobial against *A. fumigatus* and *Candida albicans*	Wang et al. (2012)
		Dihydroaltenuene A **(40)**		
Cladonia ochrochlora	*Phialophora* sp.	Xinshengin **(41)**[a]		Ye et al. (2013)
		Phialophoriol **(42)**[a]		
Parmelinella wallichiana	*Nigrospora* sp.	Altenuene **(43)**		Wang et al. (2012), He et al. (2012)
Usnea aciculifera	*Alternaria* sp.			
Cetrelia braunsiana	*Phialophora* sp.			
Hypotrachyna (= *Everniastrum* sp.)	*Alternaria* (= *Ulocladium* sp.)			
Parmelinella wallichiana	*Nigrospora* sp.	(+)-(2S,3S,4aS)-Altenuene **(43a)**[a]		He et al. (2012)
Usnea aciculifera	*Alternaria* sp.			
Cetrelia braunsiana	*Phialophora* sp.			
Parmelinella wallichiana	*Nigrospora* sp.	(−)-(2R,3R,4aR)-Altenuene **(43b)**		
Usnea aciculifera	*Alternaria* sp.			
Cetrelia braunsiana	*Phialophora* sp.			
Parmelinella wallichiana	*Nigrospora* sp.	Isoaltenuene **(44)**	Antimicrobial against *B. subtilis*	Wang et al. (2012), He et al. (2012)
Usnea aciculifera	*Alternaria* sp.			
Cetrelia braunsiana	*Phialophora* sp.			
Hypotrachyna (= *Everniastrum* sp.)	*Alternaria* (= *Ulocladium* sp.)			

Parmelinella wallichiana	*Nigrospora* sp.	(−)-(2S,3S,4aR)-Isoaltenuene (**44a**)[a]		He et al. (2012)
Usnea aciculifera	*Alternaria* sp.			
Cetrelia braunsiana	*Phialophora* sp.			
Parmelinella wallichiana	*Nigrospora* sp.	(+)-(2R,3R,4aS)-Isoaltenuene (**44b**)		
Usnea aciculifera	*Alternaria* sp.			
Cetrelia braunsiana	*Phialophora* sp.			
Parmelinella wallichiana	*Nigrospora* sp.	Alternariol (**45**)	Antiviral activity against herpes simplex virus	He et al. (2012)
Usnea aciculifera	*Alternaria* sp.			
Cetrelia braunsiana	*Phialophora* sp.			
Parmelinella wallichiana	*Nigrospora* sp.	Alternariol-9-methyl ether (**46**)		
Usnea aciculifera	*Alternaria* sp.			
Cetrelia braunsiana	*Phialophora* sp.			
Parmelinella wallichiana	*Nigrospora* sp.	4-Hydroxyalternariol-9-methyl ether (**47**)		
Usnea aciculifera	*Alternaria* sp.			
Cetrelia braunsiana	*Phialophora* sp.			
Hypotrachyna (= *Everniastrum* sp.)	*Alternaria* (= *Ulocladium* sp.)	7-Hydroxy-3,5-dimethyl-isochromen-1-one (**48**)[a]	Antimicrobial against *C. albicans*	Wang et al. (2012)
		6-Hydroxy-8-methoxy-3a-methyl-3a,9b-dihydro-3H furo[3,2-c] isochromene-2,5-dione (**49**)[a]	Antimicrobial against *B. subtilis*	
		7-Hydroxy-3-(2-hydroxy-propyl)-5-methyl isochromen-1-one (**50**)	Antimicrobial against *C. albicans*	

(continued)

Table 1 (continued)

Host	Endolichenic fungi	Secondary metabolites	Bioactive properties	References
Zeroviella mandschurica (= *Xanthoria mandschurica*)	*Coniochaeta* sp.	Conioxanthone A (**51**)[a]		Wang et al. (2010a)
		Microxanthone (**52**)		
		Moniliphenone (**53**)		
		Isosulochrin (**54**)		
Nephromopsis pallescens	*Sporormiella minima*	Sporormiellone A (**55**)[a]		Xiong et al. (2014)
		Sporormiellone B (**56**)		
Hypotrachyna (= *Everniastrum* sp.)	*Alternaria* (= *Ulocladium* sp.)	Norlichexanthone (**57**)	Antimicrobial against *A. fumigatus*, *B. subtilis* and methicillin-resistant *Staphylococcus aureus*	Wang et al. (2012)
		Griseoxanthone C (**58**)	Antimicrobial against *B. subtilis* and *C. albicans*	
		6-O-Methylnorlichexanthone (**59**)	Antimicrobial against *B. subtilis*	
Ramalina calicaris	*Preussia africana*	2,8-Dihydroxy-3-methyl-9-oxoxanthene-1-carboxylic acid methylester (**60**)		Zhang et al. (2012)
Zeroviella mandschurica (= *Xanthoria mandschurica*)	*Coniochaeta* sp.	Conioxepinol A (**61**)[a]	Cytotoxic against HeLa cell line	Wang et al. (2010b)
		Conioxepinol B (**62**)[a]		
		Conioxepinol C (**63**)[a]	Cytotoxic against A-549 and MDA-MB-231 cell line	
		Conioxepinol D (**64**)[a]		
		Brocaenol A (**65**)		
		Coniofurol (**66**)		
Cetraria islandica	*Myxotrichum* sp.	Myxotrichin A (**67**)[a]		Yuan et al. (2013)
		Myxotrichin B (**68**)[a]		
		Myxotrichin C (**69**)[a]		

Nephromopsis pallescens	*Sporormiella minima*	Sporormiellin A **(70)**[a]		Xiong et al. (2014)
		Sporormiellin B **(71)**[a]		
		Sporormiellin C **(72)**[a]		
		Microsphaeropsone A **(73)**		
Zeroviella mandschurica (= *Xanthoria mandschurica*)	*Coniochaeta* sp.	Coniothiepinol A **(74)**[a]	Antimicrobial against *Enterococcus faecalis, E. faecium* and *Fusarium oxysporum*	Wang et al. (2010b)
		Coniothiepinol B **(75)**[a]		
		Coniothienol A **(76)**[a]	Antimicrobial against *E. faecalis* and *E. faecium*	
Ramalina calicaris	*Preussia africana*	Preussochromone A **(77)**[a]	Cytotoxic against A-549, HeLa and HCT-116 cell line	Zhang et al. (2012)
Peltigera elisabethae (= *Peltigera elisabethae* var. *mauritzi*)	*Aspergillus* sp.	Oxisterigmatocystin D **(78)**[a]		Zhao et al. (2014)
		Oxisterigmatocystin A **(79)**		
		Oxisterigmatocystin C **(80)**		
		Sterigmatocystin **(81)**		
Hypotrachyna cirrhata (= *Evernistrum cirrhatum*)	*Chaetomium elatum*	Xanthoquinodin A4 **(82)**[a]	Cytotoxic against SMMC-7721, A-549, MCF-7 and SW-480 cell line	Chen et al. (2013)
		Xanthoquinodin A5 **(83)**[a]	Cytotoxic against HL-60, SMMC-7721, A-549, MCF-7 and SW-480 cell line	
		Xanthoquinodin A6 **(84)**[a]		
		Xanthoquinodin B4 **(85)**[a]	Cytotoxic against HL-60, SMMC-7721, A-549 and MCF-7 cell line	
		Xanthoquinodin B5 **(86)**[a]	Cytotoxic against HL-60, SMMC-7721, A-549, MCF-7 and SW-480 cell line	
		Xanthoquinodin A1 **(87)**		
		Xanthoquinodin A2 **(88)**	Cytotoxic against HL-60, SMMC-7721, A-549 and MCF-7 cell line	
		Xanthoquinodin A3 **(89)**	Cytotoxic against HL-60, SMMC-7721, A-549, MCF-7 and SW-480 cell line	

(continued)

Table 1 (continued)

Host	Endolichenic fungi	Secondary metabolites	Bioactive properties	References
Heterodermia obscurata	*Phaeosphaeria* sp.	Phaeosphaerin A (**90**)[a]	Cytotoxic against PC-3, DU-145 and LNCaP cell line	Li et al. (2012a, b)
		Phaeosphaerin B (**91**)[a]		
		Phaeosphaerin C (**92**)[a]	Cytotoxic against PC-3, DU-145, LNCaP, K-562 (with light) and K-562 (without light) cell line	
		Phaeosphaerin D (**93**)[a]	Cytotoxic against PC-3, DU-145 and LNCaP cell line	
		Phaeosphaerin E (**94**)[a]		
		Phaeosphaerin F (**95**)[a]		
		Hypocrellin A (**96**)	Cytotoxic against K-562 (with light) and K-562 (without light) cell lines	
		Hypocrellin C (**97**)		
		Elsinochrome A (**98**)		
		Elsinochrome B (**99**)		
		Elsinochrome C (**100**)		
		(+)-Calphostin D (**101**)		
Xanthoparmelia sp.	*Peziza* sp.	Pezizolide A (**102**)[a]		Zhang et al. (2014)
		Pezizolide B (**103**)[a]		
		Pezizolide C (**104**)[a]		
		Pezizolide D (**105**)[a]		
		Pezizolide E (**106**)[a]		
		Pezizolide F (**107**)[a]		
		Pezizolide G (**108**)[a]		

Clavaroids sp.	*Pestalotiopsis* sp.	Ambuic acid (**109**)	Antimicrobial against gram-positive bacterium *S. aureus*	Ding et al. (2009)
		Ambuic acid (**110**)[a]		
		Ambuic acid (**111**)[a]		
		Ambuic acid (**112**)[a]		
		Ambuic acid (**113**)[a]		
		Ambuic acid (**114**)[a]		
		Ambuic acid (**115**)[a]		
Cetraria islandica	*Myxotrichum* sp.	Myxodiol A (**116**)[a]	Cytotoxic against K-562 cell line	Yuan et al. (2013)
			Antimicrobial against *C. albicans*	
Ramalina calicaris	*Chaetomium elatum*	Chaetomugilin S (**117**)[a]	Caspase-3 inhibitory activity	Chen et al. (2012)
		7,50-bis-epi-Chaetoviridin A (**118**)[a]		
		7-epi-Chaetoviridin E (**119**)[a]		
Clavaroids sp.	*Pestalotiopsis* sp.	7-epi-Chaetoviridin E (**120**)[a]		Ding et al. (2009)
Hypotrachyna (= *Everniastrum* sp.)	*Alternaria* (= *Ulocladium* sp.)	TCA F (**121**)[a]		Wang et al. (2013a)
		TCA G (**122**)[a]		
		TCA H (**123**)[a]		
		TCA 1a (**124**)		
		TCA 1b (**125**)	Antimicrobial against Bacillus Calmette-Guérin (BCG)	
		TCA 2a (**126**)		
		TCA 2b (**127**)		
		TCA 6a (**128**)		
		TCA 6b (**129**)		
		TCA 9b (**130**)	Cytotoxic against HeLa cell line	
		Tricycloalternarenal (**131**)		
		ACTG-toxin H (**132**)		
		TCAD (**133**)		

(continued)

Table 1 (continued)

Host	Endolichenic fungi	Secondary metabolites	Bioactive properties	References
Pseudevernia intensa	*Geopyxis aff. Majilis*	Geopyxin A (**134**)[a]		Wijeratne et al. (2012)
		Geopyxin B (**135**)[a]	Cytotoxic against NCI-H460, SF-268, MCF-7, PC-3 M and MDA-MB-231 cell lines	
		Geopyxin C (**136**)[a]		
		Geopyxin D (**137**)[a]		
	Geopyxis sp.	Geopyxin E (**138**)[a]		
	Geopyxis sp.	Geopyxin F (**139**)[a]		
Hypotrachyna (= *Everniastrum* sp.)	*Alternaria* (= *Ulocladium* sp.)	Ophiobolin P (**140**)[a]	Cytotoxic against KB and HepG2 cell lines	Wang et al. (2013b)
			Antimicrobial against *B. subtilis* and methicillin-resistant *S. aureus*	
		Ophiobolin Q (**141**)[a]	Cytotoxic against KB and HepG2 cell lines	
		Ophiobolin R (**142**)[a]		
		Ophiobolin S (**143**)[a]		
		Ophiobolin T (**144**)[a]		
			Antimicrobial against *A. fumigatus*, *B. subtilis*, methicillin-resistant *S. aureus* and Bacillus Calmette-Guérin	
		6-epi-21,21-Odihydroophiobolin G (**145**)	Cytotoxic against KB and HepG2 cell lines	
			Antimicrobial against Bacillus Calmette-Guérin	
		6-epi-Ophiobolin G (**146**)	Cytotoxic against KB and HepG2 cell lines	
			Antimicrobial against Bacillus Calmette-Guérin	
		6-epi-Ophiobolin K (**147**)	Cytotoxic against KB and HepG2 cell lines	
			Antimicrobial against *B. subtilis* and methicillin-resistant *S. aureus*	

Usnea mutabilis	*Sporormiella irregularis*	22E-3,7-Epoxy-5,10:8,9-disecoergosta-9(10),22-diene-5,8-dione **(148)**[a]		Li et al. (2012a, b)
		22E-5a,6a-Epoxyergosta-8(14),22-diene-3b,7a-diol **(149)**		
		22E-5a,6a-Epoxyergosta-8(9),22-diene-3b,7a-diol **(150)**		
		22E-7a-Methoxy-5a,6a-epoxyergosta-8(14),22-dien-3b-ol **(151)**		
		22E-3b-Hydroxy-5a,6a-epoxyergosta-22-en-7-one **(152)**		
		22E-Ergosta-7,22-diene-3b,5a,6b-triol **(153)**		
		22E-6b-Methoxyergosta-7,22-diene-3b,5a-diol **(154)**		
		22E-5a,8a-Epidioxyergosta-6,22-dien-3b-ol **(155)**		
		22E-Ergosta-4,6,8(14),22-tetraen-3-one **(156)**		
Hypotrachyna (= *Everniastrum* sp.)	*Nodulisporium* sp.	Inoterpene B **(157)**		Zheng et al. (2013)
		Nodulisporisteroid A **(158)**[a]		
		Nodulisporisteroid A **(159)**[a]	$A\beta_{42}$ aggregation inhibitory activity	
		Demethoxyviridin **(160)**		

(continued)

Table 1 (continued)

Host	Endolichenic fungi	Secondary metabolites	Bioactive properties	References
Cladina grisea	*Eurotium* sp.	Neoechimulin A **(161)**		Chen et al. (2014)
		Isoechinulin A **(162)**		
		Tardioxopiperazine A **(163)**		
		Dehydroechinulin **(164)**		
		7-O-Methylvariecolortide A **(165)**[a]	Caspase-3 inhibitory activity	
		(−)-(S)-7-O-Methylvariecolortide A **(165a)**[a]		
		(+)-(R)-7-O-Methylvariecolortide A **(165b)**[a]		
		Variecolortide B **(166)**[a]	Caspase-3 inhibitory activity	
		(−)-(S)-Variecolortide B **(166a)**[a]		
		(+)-(R)-Variecolortide B **(166b)**[a]		
		Variecolortide C **(167)**[a]	Caspase-3 inhibitory activity	
		(−)-(S)-Variecolortide C **(167a)**[a]		
		(+)-(R)-Variecolortide C **(167b)**[a]		
Hypotrachyna nepalensis (= *Everniastrum nepalense*)	*Chaetomium globosum*	Chaetoglobosin Y **(168)**[a]		Zheng et al. (2014)
		Chaetoglobosin F **(169)**		
		Chaetoglobosin E **(170)**	Cytotoxic against HCT-116 cell line	
		Isochaetoglobosin D **(171)**	Cytotoxic against HCT-116 cell line	
		Chaetoglobosin G **(172)**		
		Chaetoglobosin B **(173)**		
		Chaetoglobosin C **(174)**	Cytotoxic against HCT-116 cell line	

Leptogium saturninum	*Xylaria* sp.	Cyclo(N-methyl-L-Phe-L-Val-D-Ile-L-Leu-L-Pro) (**175**)[a]	Antimicrobial against *C. albicans*	Wu et al. (2011)
		Cyclo(L-Val-D-Ile-L-Leu-L-pro-D-Leu) (**176**)[a]		
Usnea mutabilis	*Sporormiella irregularis*	Sporormielloside (**177**)[a]		Yang et al. (2016)
		Calyxanthone-8-methyl ether (**178**)		
		Endocrocin (**179**)		
Lethariella zahlbruckneri	*Tolypocladium cylindrosporum*	Tolypocladenols A1 (**180**)[a]		XB Li et al. (2015b)
		Tolypocladenols A2 (**181**)[a]		
		Tolypocladenols B (**182**)[a]		
		Tolypyridone A (**183**)		
		3,8-dihydroxy-4-(4-hydroxyphenyl)-6-methylcoumarin (**184**)		
		F-14329 (**185**)		
		Pyridoxatin (**186**)	Cytotoxic against MDA-MB-231, A2780, K562, and A549 cancer cell lines	
		Tolypyridone B (**187**)		
		Terpendole E (**188**)	Cytotoxic against MDA-MB-231, A2780, K562 and A549 cancer cell lines	
Cetraria islandica	*Myxotrichum* sp.	7,8-dihydro-7*R*,8*S*-dihydroxy-3,7-dimethyl-2-benzopyran-6-one (**189**)[a]		Yuan et al. (2016)
		Myxotritones A (**190**)[a]		
		Myxotritones B (**191**)[a]		
		Myxotritones C (**192**)[a]		

(continued)

Table 1 (continued)

Host	Endolichenic fungi	Secondary metabolites	Bioactive properties	References
Parmelia sp.	*Nectria* sp.	Necpyrones A (**193**)[a]		G Li et al. (2015a)
		Necpyrones B (**194**)[a]		
		Necpyrones C (**195**)[a]		
		Necpyrones D (**196**)[a]		
		Necpyrones E (**197**)[a]		
		PC-2 (**198**)		
		(6S, 1′*S*)-LL-P880α (**199**)		
		(6S, 1′*S*, 2′*R*)-LL-P880β (**200**)		
		(1′*S*, 2′*R*)-1-hydroxy-1-((*S*)-4-methoxy-6-oxo-3,6-dihydro-2H-pyran-2-yl)-pentan-2-yl acetate (**201**)		
		(*S*)-4-methoxy-6-pentanoyl-5,6-dihydro-2H-pyran-2-one (**202**)		
		(1′*S*, 2′*R*)-LL-P880γ (**203**)		
		Helvolic acid (**204**)		
Lethariella zahlbruckneri	*Tolypocladium cylindrosporum*	Pyridoxatin (**205**)		Chang et al. (2015)
		N-deoxy-pyridoxatin (**206**)	Antimicrobial against *C. albicans*	

Host	Endolichenic fungi	Secondary metabolites	Bioactive properties	References
Pseudocyphellaria sp.	*Biatriospora* sp.	Biatriosporin A (**207**)[a]		Zhou et al. (2016a)
		Biatriosporin B (**208**)[a]		
		Biatriosporin C (**209**)[a]		
		Biatriosporin D (**210**)[a]	Antimicrobial against *C. albicans*	
		Biatriosporin E (**211**)[a]		
		Biatriosporin F (**212**)[a]		
		Biatriosporin G (**213**)[a]		
		Biatriosporin H (**214**)[a]		
		Biatriosporin I (**215**)[a]		
		Biatriosporin J (**216**)[a]		
		Biatriosporin K (**217**)[a]		
		Biatriosporin L (**218**)[a]		
		Biatriosporin M (**219**)[a]		
		Balticol A (**220**)		
		Scytalols D (**221**)		
		Balticol E (**222**)		

(continued)

Table 1 (continued)

Host	Endolichenic fungi	Secondary metabolites	Bioactive properties	References
		6-O-demethyl-5-deoxyfusarubin (**223**)		
		Pyranonaphthoquinone (**224**)		
		Balticol F (**225**)		
		Scytalols C (**226**)		
		6-deoxy-7-O-demethyl-3,4-anhydrofusarubin (**227**)		
		6-deoxy-3,4-anhydrofusarubin (**228**)		
		Ascomycone A (**229**)		
		ZSU-H85 A (**230**)		
		3-acetyl-2,8-dihydroxy-6-methoxy anthraquinone (**231**)		
		Pleorubrin B (**232**)		
		6-deoxybostrycoidin (**233**)		
		2-acetonyl-3-methyl-5-hydroxy-7-methoxynaphthazarin (**234**)		
		5-hydroxymellein (**235**)		
		4-hydroxymellein (**236**)		
		3,4-dihydro-4,5,8-trihydroxy-3-methylisocoumarin (**237**)		
		5-hydroxyramulosin (**238**)		

Lobaria quercizans	*Aspergillus versicolor*	Diorcinols F (**239**)[a]		XB Li et al. (2015b)
		Diorcinols G (**240**)[a]	Cytotoxic against PC3, A549, A2780, MDA-MB-231, and HEPG2 cancer cell lines	
		Diorcinols H (**241**)[a]		
		3-methoxyviolaceol-II (**242**)[a]		
		(−)-(R)-cyclo-hydroxysydonic acid (**243**)[a]		
		(−)-(7S,8R)-8-hydroxysydowic acid (**244**)[a]		
		(−)-(7R,10S)-10-hydroxysydowic acid (**245**)[a]		
		(−)-(7R,10R)-iso-10-hydroxysydowic acid (**246**)[a]		
		(−)-12-acetoxy-1-deoxysydonic acid (**247**)[a]		
		(−)-12-acetoxysydonic acid (**248**)[a]		
		(−)-12-hydroxysydonic acid (**249**)[a]		
		(−)-(R)-11-dehydrosydonic acid (**250**)[a]		
		Sydowiol D (**251**)[a]		
		Sydowiol E (**252**)[a]		
		Diorcinol I (**253**)	Cytotoxic against PC3, A549, A2780, MDA-MB-231, and HEPG2 cancer cell lines	
			Antimicrobial against *C. albicans*	
		Diorcinol (**254**)		
		Diorcinol D (**255**)	Cytotoxic against PC3, A549, A2780, MDA-MB-231, and HEPG2 cancer cell lines	
		Violaceol I (**256**)	Antimicrobial against *C. albicans*	
		Cordyol C (**257**)		

(continued)

Table 1 (continued)

Host	Endolichenic fungi	Secondary metabolites	Bioactive properties	References
		3,7-dihydroxy-1,9-dimethyldibenzofuran (**258**)	Cytotoxic against PC3, A549, A2780, MDA-MB-231, and HEPG2 cancer cell lines	
			Antimicrobial against *C. albicans*	
		Violaceol II (**259**)	Antimicrobial against *C. albicans*	
		Sydowic acid (**260**)		
		Sydonic acid (**261**)		
		Sydowiol A (**262**)		
		Sydowiol B (**263**)		
		Aversin (**264**)		
		6,8-di-O-methylnidurufin (**265**)		
		4-hydroxybenzaldehyde (**266**)		
		3-formyl-1H-indole (**267**)		
Lobaria retigera	*Aspergillus versicolor*	8-*O*-methylversicolorin B (**268**)[a]	Cytotoxic against PC-3 and H460 cancer cell lines	Dou et al. (2014)
		8-*O*-methylversicolorin A (**269**)[a]		
		8-*O*-methylaverythin (**270**) [a]		
		1′-*O*-ethyl-6,8-di-*O*-methylaverantin (**271**)[a]		

Sticta fuliginosa	Xylariaceae fungal strain CR1546C	(R)-4,6,8-trihydroxy-3,4-dihydro-1(2H)-naphthalenone (**272**)[a]	Antimicrobial against *B. subtilis* and *C. albicans*	Kim et al. (2014)
		6,8-dihydroxy-(3R)-(2-oxopropyl)-3,4-dihydroisocoumarin (**273**)[a]		
		6,8-dihydroxy-(3)-(2-oxopropyl)-isocoumarin (**274**)	Antimicrobial against *C. albicans*	
		6,8-dihydroxy-3-[(2*S*)-2-hydroxypropyl]-isocoumarin (**275**)		
		2,4-dihydroxy-6-(2-oxopropyl)-benzoic acid (**276**)		
		6,8-dihydroxy-3-methylisocoumarin (**277**)		
		5,6,8-trihydroxy-3(*R*)-methyl-3,4-dihydroisocoumarin (**278**)		
		6,8-dihydroxy-3(*R*)-methyl-3,4-dihydroisocoumarin (**279**)		
		(3*R*,4*S*)-3,4,8-trihydroxy-3,4-dihydro-1(2*H*)-naphthalenone (**280**)		
		(3*S*,4*S*)-3,4,6,8-tetrahydroxy-3,4-dihydro-1(2*H*)-naphthalenone (**281**)		
Parmelia sp.	*Periconia* sp.	Pericoterpenoid A (**282**)[a]	Antimicrobial against *A. niger* and *C. albicans*	Wu et al. (2015)
Hypotrachyna (= *Everniastrum* sp.)	*Hypoxylon* (= *Nodulisporium* sp.)	Nodulisporipyrones A (**283**)[a]	Antimicrobial against *A. niger* and *C. albicans*	Q Zhao et al. (2015b)
		Nodulisporipyrones B (**284**)[a]		
		Nodulisporipyrones C (**285**)[a]		
		Nodulisporipyrones D (**286**)[a]		

(continued)

Table 1 (continued)

Host	Endolichenic fungi	Secondary metabolites	Bioactive properties	References
Hypotrachyna (= *Everniastrum* sp.)	*Hypoxylon* (= *Nodulisporium* sp.)	Nodulisporisteroids C (**287**)[a]		Q Zhao et al. (2015c)
		Nodulisporisteroids D (**288**)[a]		
		Nodulisporisteroids E (**289**)[a]		
		Nodulisporisteroids F (**290**)[a]		
		Nodulisporisteroids G (**291**)[a]		
		Nodulisporisteroids H (**292**)[a]		
		Nodulisporisteroids I (**293**)[a]		
		Nodulisporisteroids J (**294**)[a]		
		Nodulisporisteroids K (**295**)[a]		
		Nodulisporisteroids L (**296**)[a]		
Hypotrachyna (= *Everniastrum* sp.)	*Hypoxylon* (= *Nodulisporium* sp.)	Nodulisporiviridins A (**297**)[a]	$A\beta_{42}$ aggregation inhibitory activity	Q Zhao et al. (2015a)
		Nodulisporiviridins B (**298**)[a]		
		Nodulisporiviridins C (**299**)[a]		
		Nodulisporiviridins D (**300**)[a]		
		Nodulisporiviridins E (**301**)[a]		
		Nodulisporiviridins F (**302**)[a]		
		Nodulisporiviridins G (**303**)[a]		
		Nodulisporiviridins H (**304**)[a]		

Unknown	Pleosporales sp.	Cucurbitarin A (**305**)[a]		Jiao et al. (2015)
		Cucurbitarin B (**306**)[a]		
		Cucurbitarin C (**307**)[a]		
		Cucurbitarin D (**308**)[a]		
		Cucurbitarin E (**309**)[a]		
		3,10-dihydroxy-4,8-dimethoxy-6-methylbenzocoumarin (**310**)[a]		
		3,8,10-trihydroxy-4-methoxy-6-methylbenzocoumarin (**311**)[a]		
		(5*R*)-5-hydroxy-2,3-dimethylcyclohex-2-en-1-one (**312**)[a]		
		2,5-dimethoxy-3,6-bis(4-methoxypheny1)-1,4-benzoquinone (**313**)		
		Dankasterone A (**314**)		
		(17*R*)-4-hydroxy-17-methylincisterol (**315**)		
Cladonia evansii	*Coniochaeta* (= *Lecythophora* sp. FL1031)	Oxaspirol A (**316**)		Wijeratne et al. (2016)
		Oxaspirol B (**317**)	p97 inhibitory activity	
		Oxaspirol C (**318**)		
Parmotrema tinctorum	*Coniochaeta* (= *Lecythophora* sp. FL1375)	Oxispirol B (**317**)	p97 inhibitory activity	
		Oxispirol C (**318**)		
		Oxaspirol D (**319**)[a]		

(continued)

Table 1 (continued)

Host	Endolichenic fungi	Secondary metabolites	Bioactive properties	References
Cetraria islandica	*Pestalotiopsis* sp.	Ambuic acid (**320**)[a]	Antimicrobial against *F. oxysporum*	Yuan et al. (2017)
		Ambuic acid (**321**)[a]		
		Ambuic acid (**322**)[a]		
		Ambuic acid (**323**)[a]		
		Ambuic acid (**324**)[a]	Antimicrobial against *F. oxysporum* and *F. graminum*	
		Ambuic acid (**325**)	Antimicrobial against *F. graminum*	
		Ambuic acid (**326**)		
		Ambuic acid (**327**)		
		Ambuic acid (**328**)		
		Ambuic acid (**329**)		
		Ambuic acid (**330**)		
Umbilicaria sp.	*Chaetothyriales* sp.	Chaetothyrins A (**331**)[a]		Zhou et al. (2016b)
		Chaetothyrins B (**332**)[a]		
		Chaetothyrins C (**333**)[a]	Radical scavenging activity on DPPH	
Parmotrema sp.	*Penicillium citrinum*	5′-acetyl-3,5,7′-trimethoxy-3′H-spiro[cyclohexa[2,4]diene-1,1′-isobenzofuran]-3′,6-dione (**334**)	Radical scavenging activity on DPPH	Samanthi et al. (2015)
		4-acetyl-2′-hydroxy-3′,5′,6-trimethoxybiphenyl-2-carboxylic acid (**335**)		

Leptogium hildenbrandii	*Massarina* sp.	(11S,12S,13R)-11,13-dihydroxy-12-methyltetradecanoic acid (**336**)[a]		Yuan et al. (2015)
		Cyclo-(S-Pro-R-Leu) (**337**)		
		3,4-dihydro-3,4,8-trihydroxy(2H) naphthalenone (**338**)		
Endocarpon pusillum	EL002332	Crude acetone extract	Cytotoxic against AGS human gastric cancer cells and CT26 mouse colon cancer cells; inhibited tumor growth in both skin tumors and an intraperitonealxenograft mouse model	Yang et al. (2018)
		Crude acetone extract + docetaxel	Anticancer against AGS human gastric cancer cells and TMK1 cells	
		myA		
		myB		
		myC	Cytotoxic against AGS human gastric cancer cells	
Stereocaulon tomentosum	Dothideomycetes sp. EL003334	Dothideopyrones E (**339**)[a]		GS Kim et al. (2018a)
		Dothideopyrones F (**340**)[a]	Inhibited nitric oxide (NO) production in lipopolysaccharide (LPS)-induced BV2 cells; diminished the protein expression levels of inducible nitric oxide synthase (iNOS) and cyclooxygenase-2 (COX-2); decreased the mRNA expression levels of pro-inflammatory cytokines	

(continued)

Table 1 (continued)

Host	Endolichenic fungi	Secondary metabolites	Bioactive properties	References
Unknown	*Phoma* sp.	Phomalichenones A **(341)**[a]	Inhibited nitric oxide (NO) production in lipopolysaccharide (LPS)-stimulated RAW264.7 macrophages; diminished the protein expression levels of inducible nitric oxide synthase (iNOS) and cyclooxygenase-2 (COX-2); decreased the mRNA expression levels of pro-inflammatory cytokines	JW Kim et al. (2018b)
		Phomalichenones B **(342)**[a]		
		Phomalichenones C **(343)**[a]		
		Phomalichenones D **(344)**[a]		
		(2,4-dihydroxy-3-(2-hydroxyethyl)-6-methoxyphenyl)-3 hydroxybutan-1-one **(345)**		
		(E)-1-(2,4- dihydroxy-3-(2-hydroxyethyl)-6-methoxyphenyl) but-2-en-1-one **(346)**	Inhibited nitric oxide (NO) production in lipopolysaccharide (LPS)-stimulated RAW264.7 macrophages	
		Phomalone **(347)**		
		Deoxyphomalone **(348)**		
		8-ethyl-7-hydroxy-5-methoxy-2-methylchroman-4-one **(349)**		
		LL-D253γ **(350)**		
		4-hydroxy-6-methoxy-5-(1′-oxobutyl) benzodihydrofuran **(351)**		

[a] new compounds

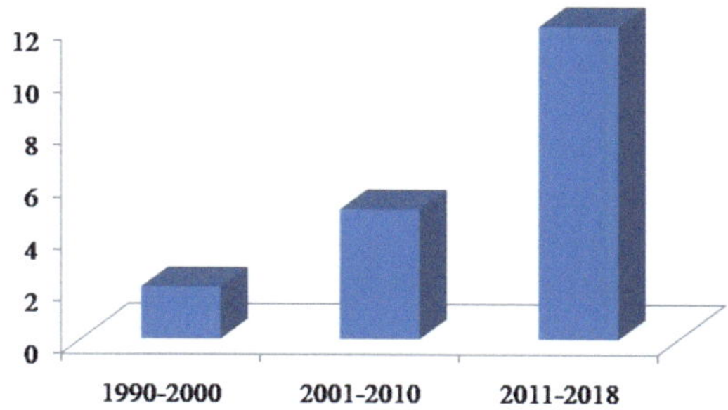

Fig. 1 Decade-wise gradual increase in research publications dealing with secondary metabolites isolated from endolichenic fungi and their bioactive potential

Five novel polyketides [Globosumoside A (**13**) and B (**12**), Chaetomugilin S (**117**), 7,50-bis-epi-Chaetoviridin A (**118**), 7-epi-Chaetoviridin E (**119**)] were reported from *Chaetomium elatum* isolated from *Ramalina calicaris* and three of which (**117–119**) exhibited caspase-3 inhibitory activity (Chen et al. 2012).

The ethyl acetate extracts of *Nigrospora sphaerica* from *Parmelinella wallichiana*, *Alternaria alternata* from *Usnea aciculifera* and *Phialophora* sp. from *Cetrelia braunsiana* yielded Alternariol (**45**) and Alternariol-9-methyl ether (**46**) having antiviral activity against Herpes Simplex Virus (He et al. 2012).

Huang et al. (2012) reported the occurrence of 2-Isopentenyldiorcinol (**4**) and Diorcinol (**5**) in *Aspergillus* sp. isolated from *Peltigera elisabethae.*

An endolichenic fungus *Phaeosphaeria* sp., isolated from *Heterodermia obscurata*, found possessing six novel polyketides [Phaeosphaerin A-F (**90**)] showed cytotoxic potential against PC-3, DU-145, LNCaP, K-562 cancer cell lines (Li et al. 2012b).

Li et al. (2012a) extracted a novel steroid 22E-3,7-Epoxy-5,10:8,9-disecoergosta-9(10),22-diene-5,8-dione (**148**) from an endolichenic fungus *Sprormiella irregularis* isolated from *Usnea mutabilis*.

Two novel polyketides *viz.* 1,8-Dimethoxynaphthlen-2-ol (**10**) and 1-(40-Hydroxyl-30,50-dimethoxy-phenyl)-1,8-dimethoxynaphthalen-2(1H)-one (**11**) were acquired from *Scopulariopsis* sp. isolated from *Cladonia gracilis* (Yang et al. 2012).

Wang et al. (2012) reported various metabolites including two novel compounds from cultures of *Ulocladium* sp. isolated from *Evernistrum* sp., which showed radical scavenging activity on DPPH [5′-Methoxy-6-methyl-biphenyl-3,4,3′-triol (**1**), Altenusin (**2**), Alterlactone (**3**)] and antimicrobial activity against *Aspergillus fumigatus*, *Bacillus subtilis*, *Candida albicans*, methicillin-resistant *Staphylococcus aureus* [Rubralactone (**39**), Isoaltenuene (**44**), 7-Hydroxy-3,5-dimethyl-isochromen-1-one (**48**), 6-Hydroxy-8-methoxy-3a-methyl-3a,9b-dihydro-3H furo[3,2-c]isochromene-2,5-dione (**49**), 7-Hydroxy-3-(2-hydroxy-propyl)-5-methyl isochromen-1-one (**50**), Norlichexanthone (**57**), Griseoxanthone C (**58**), 6-O-Methylnorlichexanthone (**59**)].

Wijeratne et al. (2012) obtained six novel ent-kaurane terpenoids [Geopyxin A-F (**134–139**) from *Geopyxis* aff. *Majalis* and *Geopyxis* sp. isolated from *Pseudevernia intensa*. Geopyxin B revealed to be the only natural geopyxin exhibiting cytotoxic activity against the cancer cell lines NCI-H460, SF-268, MCF-7, PC-3M, and MDA-MB-231.

Six novel preussochromone derivatives [Preussochromone A-F (**34–38,77**)] were mined from *Preussia africana* isolated from *Ramalina calicaris* by Zhang et al. (2012). Among these six compounds, preussochromone A (**77**) and preussochromone C (**35**) demonstrated significant cytotoxic activity against A-549, HeLa and HCT-116 cancer cell lines.

Investigation of endolichenic fungus *Chaetomium elatum* isolated from *Everniastrum cirrhatum*, revealed eight xanthoquinodins (**82–89**), of which five were novel (**82–86**). All of these possess cytotoxic activity against five cancer cell lines (HL-60, SMMC-7721, A-549, MCF-7 and SW480) (Chen et al. 2013).

Wang et al. (2013a, b) reported eight novel terpenoids [TCA F (**121**), TCA G (**122**), TCA H (**123**), Ophiobolin P-T (**140–144**)] from *Ulocladium* sp. isolated from *Everniastrum* sp., of which TCA 1b (**125**) showed antimicrobial efficacy against bacilli *Calmette-Guerin*; TCA 9b (**130**) was cytotoxic against HeLa tumor cell line; Ophiobolin P (**140**) was possessings both antimicrobial efficacy against *Bacillus subtilis* and methicillin resistant- *Staphylococcus aureus* and cytotoxic effects against KB and HepG2 cancer cell lines. Ophiobolin Q, R, S (**141–143**) were found having cytotoxicity against KB and HepG2 cancer cell lines, while Ophiobolin T (**144**) was cytotoxic against KB and HepG2 cancer cell lines and also showed antimicrobial activity against *Aspergillus fumigatus*, *Bacillus subtilis*, methicillin resistant- *Staphylococcus aureus* and *Calmette-Guerin.* Other ophiobolin derivatives 6-epi-21,21-Odihydroophiobolin G (**145**), 6-epi-Ophiobolin G (**146**), 6-epi-Ophiobolin K (**147**) also showed both cytotoxic activity against KB and HepG2 cancer cell lines and antimicrobial potential against *Calmette–Guerin*, *Bacillus subtilis* and methicillin resistant- *Staphylococcus aureus.*

Altenusin (**2**) and alterlactone (**3**) were isolated from *Ulocladium* sp. harboring *Everniastrum* sp. (Wang et al. 2012).

Ye et al. (2013) isolated two polyketides *viz.* xinshengin (**41**) and phialophoriol (**42**) from *Phialophora* sp. recovered from *Cladonia ochrochlora.*

Yuan et al. (2013) isolated Myxotrichin A-C (**67–69**) extracted from *Myxotrichum* sp. from *Cetraria islandica* which demonstrated weak *in-vitro* cytotoxic activity against the K-562 tumor cell line. Along with Myxotrichin A-C, they also isolated Myxidiol A (**116**) which was quite remarkable since it is the first endolichenic fungal metabolite having a halogen atom i.e. chlorine, located at position five. This compound demonstrated antifungal activity against *Candida albicans* and cytotoxicity against K-562 cancer cell lines.

Zheng et al. (2013) gained two novel steroid derivatives [Nodulisporisteroid A (**158, 159**)] along with Inoterpene B (**157**) and Demethoxyviridin (**160**) from *Nodulisporium* sp. isolated from *Everniastrum* sp. Nodulisporisteroid A (**159**) and Demethoxyviridin (**160**) demonstrated $A\beta_{42}$ aggregation inhibitory activity.

Three novel alkaloid derivatives [7-O-Methylvariecolortide A (**165**), Variecolortide B (**166**) and Variecolortide C (**167**)] obtained from *Eurotium* sp. isolated from *Cladina grisea* showed caspase-3 inhibitory activity (Chen et al. 2014).

Investigations into the endolichenic fungal strain *Aspergillus versicolor*, isolated from *Lobaria retigera*, yielded multiple anthraquinone derivatives, including four novel and nine known derivatives, of which 8-*O*-methylversicolorin B (**268**) and 8-*O*-methylversicolorin A (**269**) showed proliferation inhibition of cancer cell lines PC-3 and H460 (Dou et al. 2014).

Two new along with eight known metabolites were obtained from a endolichenic species of Xylariaceae (CR1546C) isolated from the lichen *Sticta fuliginosa* by Kim et al. (2014). Of these ten metabolites, (R)-4,6,8-trihydroxy-3,4-dihydro-1(2H)-naphthalenone (**272**) and 6,8-dihydroxy-(3R)-(2-oxopropyl)-3,4-dihydroisocoumarin (**273**) demonstrated antimicrobial activity against *Bacillus subtilis* and *Candida albicans* while rest shows activity against *Candida albicans* only.

Xiong et al. (2014) obtained three novel secondary metabolites [Sporormiellin A-C (**70–72**)] from the cultures of an endolichenic fungus *Sporormiellla minima* isolated from *Nephromopsis pallescens*.

A novel secondary metabolite [oxisterigmatocystin D (**78**)] was isolated from *Aspergillus* sp. associated with *Peltigera elisabethae* var. *mauritzi* (Zhao et al. 2014) and diorcinol (**5**), violaceol-I and II (**6, 7**) were acquired from *Aspergillus* sp. growing in *Peltigera eilsabethae* showing Aβ_{42} aggregation inhibitory activity (Zhao et al. 2014).

Zhang et al. (2014) extracted seven novel polyketides [Pezizolide A-G (**102–108**)] from an endolichenic fungal strain *Peziza* sp. isolated from lichen *Xanthoparmelia* sp.

Four alkaloid derivatives [Chaetoglobosin E (**170**), Isochaetoglobosin D (**171**), Chaetoglobosin G (**172**) and Chaetoglobosin C (**174**)] having cytotoxic activity against HCT-116 tumor cell lines were isolated from *Chaetomium globosum* obtained from *Evernistrum nepalense* (Zheng et al. 2014).

Eight novel secondary metabolites [Cucurbitarin A (**305**), Cucurbitarin B (**306**), Cucurbitarin C (**307**), Cucurbitarin D (**308**), Cucurbitarin E (**309**), 3,10-dihydroxy-4,8-dimethoxy-6-methylbenzocoumarin (**310**), 3,8,10-trihydroxy-4-methoxy-6-methylbenzocoumarin (**311**) and (5R)-5-hydroxy-2,3-dimethylcyclohex-2-en-1-one (**312**)] were extracted from member of *Pleosporales* sp. isolated from an unknown lichen species by Jiao et al. (2015).

Five novel necpyrone derivatives [Necpyrone A-E (**193–197**)] were detected from *Nectria* sp. inhabiting *Pamelia* sp. (Li et al. 2015a).

Endolichenic fungus *Tolypocladium cylindrosporum* isolated from *Lethariella zahlbruckneri* yielded three unique tetramic acid derivatives along with some known compounds including Pyridoxatin (**186**) and Terpendole E (**188**) which revealed cytotoxic effects against MDA-MB-231, A2780, K562, and A549 cancer cell lines (XB Li et al. 2015b).

Fourteen new secondary metabolites were isolated from *Aspergillus versicolor* inhabiting *Lobaria quercizans*. Among these various compounds some showed varying degree and type of bioactive potential, such as, Diorcinols G (**240**), Diorcinols H (**241**), Diorcinol I (**253**), Diorcinol D (**255**) and 3,7-dihydroxy-1,9-dimethyldibenzofuran (**258**) were found having cytotoxicity against PC3, A549,

A2780, MDA-MB-231, and HEPG2 cancer cell lines; Diorcinol I (**253**), Diorcinol D (**255**), Violaceol I (**256**), Cordyol C (**257**), 3,7-dihydroxy-1,9-dimethyldibenzofuran (**258**) and Violaceol II (**259**) were active against *Candida albicans* (Li et al. 2015c).

Two bioactive secondary metabolites obtained from an endolichenic fungus *Penicillium citrinum* isolated from *Parmotrema* sp. possess antioxidant potential as radical scavenging activity on DPPH (Samanthi et al. 2015).

The endolichenic fungus *Massarina* sp. isolated from *Leptogium hildenbrandii* yielded a novel fatty acid (11S,12S,13R)-11,13-dihydroxy-12-methyltetradecanoic acid (**336**) along with three known metabolites, *viz.* cyclo-(S-Pro-R-Leu) (**337**), 3,4-dihydro-3,4,8-trihydroxy(2H)naphthalenone (**338**) and 4-hydroxybenzeneethanol (Yuan et al. 2015).

Eight new viridins [Nodulisporiviridins A-H (**297–304**)] were extracted from the cultures of *Nodulisporium* sp. which was isolated from *Everniastrum* sp. All these compounds exhibited $A\beta_{42}$ aggregation inhibitory activity (Q Zhao et al. 2015a).

Four novel pyrone derivatives [Nodulisporipyrones A (**283**), Nodulisporipyrones B (**284**), Nodulisporipyrones C (**285**), Nodulisporipyrones D (**286**)] were extracted from *Nodulisporium* sp. isolated from lichen *Everniastrum* sp. showed antimicrobial activity against *Aspergillus niger* and *Candida albicans* (Zhao et al. 2015b).

Ten new steroid derivatives [Nodulisporisteroids C-L (**287–296**)] were isolated from a rice culture of endolichenic fungus *Nodulisporium* sp. obtained from lichen *Everniastrum* sp. (Zhao et al. 2015c).

Zhou et al. (2016a) successfully obtained 12 novel heptaketides [Biatriosporins A-L (**207–218**)] from fungal cultures of *Biatriospora* sp. isolated from lichen *Pseudocyphellaria* sp., of which Biatriosporin D (**210**) was found active against opportunistic fungus *Candida albicans.*

A novel xanthone glycoside [Sporormielloside (**177**)] with two other known compounds Calyxanthone-8-methyl ether (**178**) and Endocrocin (**179**) were obtained from *Sporormiella irregularis* isolated from *Usnea mutabilis* (Yang et al. 2016).

Three novel polyketides [Myxotritones A-C (**190–192**)] with a new natural product 7,8-dihydro-7*R*,8*S*-dihydroxy-3,7-dimethyl-2-benzopyran-6-one (**189**) were acquired from *Myxotrichum* sp. isolated from thallus of *Cetraria islandica* (Yuan et al. 2016).

Zhou et al. (2016b) obtained three new terpenoids [Chaetothyrins A-C (**331–333**)] from an endolichenic fungus belonging to order *Chaetothyriales* and was isolated from lichen *Umbilicaria* sp.

Eleven new polyketide-terpene hybrid secondary metabolites were obtained from *Pestalotiopsis* sp. isolated from *Cetraria islandica* of which Ambuic acid (**320–324**) was showing the antimicrobial potential against *Fusarium oxysporum* and *F. graminum* (Yuan et al. 2017).

Yang et al. (2018) reported cytotoxic activity of crude extracts of endolichenic fungus (EL002332) against gastric cancer cell lines.

GS Kim et al. (2018a) isolated two new α-pyrones *viz.* dothideopyrones E (**339**) and F (**340**) from endolichenic fungus species belonging to Dothideomycetes (EL003334). Dothideopyrones F inhibited nitric oxide (NO) production in lipopolysaccharide (LPS)-induced BV2 cells, while Dothideopyrones F diminishes the protein expression levels of inducible nitric oxide synthase (iNOS) and cyclooxygenase-2 (COX-2). Additionally, Dothideopyrones F decreased the mRNA expression levels of pro-inflammatory cytokines, such as tumor necrosis factor-α (TNF-α), interleukin (IL)-1β, and IL-6.

Four novel metabolites *viz.* phomalichenones A–D (**341–344**) along with seven known compounds (**345–351**) were screened from *Phoma* sp. (EL002650) isolated from unknown lichen species. Phomalichenones A and (E)-1-(2,4-dihydroxy-3-(2-hydroxyethyl)-6-methoxyphenyl)but-2-en-1-one inhibited nitric oxide (NO) production in lipopolysaccharide (LPS)-stimulated macrophages. In addition, phomalichenones A diminished the protein expression levels of inducible nitric oxide synthase (iNOS) and cyclooxygenase-2 (COX-2), and decreased the mRNA expression levels of pro-inflammatory cytokines, such as tumor necrosis factor-α (TNF-α), interleukin(IL)-1β, and IL-6 (JW Kim et al. 2018b).

1 Conclusion

Fungi synthesize a wide variety of biologically active secondary metabolites, and after the genome sequences analysis it is now revealed that the diversity of novel natural products isolated from fungi is far greater than estimated. Molecular and evolutionary studies are providing a novel perspective to the evolution of genes responsible for the production of secondary metabolites. Now it is well known that various fungal species comprise the genetic capacity to synthesize numerous secondary metabolites, some of which are species specific while others are of quite common occurrence. The biological function of the secondary metabolites is not yet known; though, they may be responsible for the various types of fungal lifestyles (as parasite, symbiont, saprophyte), virulence, and in defense. Cooperative efforts from various disciplines (such as bioinformatics, genetics, physiology) will enlighten the genesis and function of secondary metabolites.

Lichens which are very slow growing organisms are harnessed on industrial scale for making pharmaceuticals because of the presence of ca. 1000 secondary metabolites which are unique to them. Developing a drug from lichens on one hand not only take a lot of time and effort, but on the other hand is quite unrealistic and costly affair, as they are very slow growing creatures. In comparison to lichens, endolichenic fungi can be harnessed with ease in search for bioactive and chemically novel therapeutic agents. Studies have shown that endolichenic fungi like other endophytic fungi have a huge opportunity in acquiring novel bioactive metabolites because these fungi not only could produce similar compounds as of their respective host (Strobel and Daisy 2003) but also some other novel compounds, and because of this property they have good antimicrobial potential. If the bioactive compounds of endolichenic fungi would be offered to the pharmaceutical industries, then there will be no need to mow the lichens for bioactive metabolites. This on a mass scale can be achieved with the help of innovative and promising technologies like metabolomics, interactomics, and chemoinformatics, which will hopefully be less complicated in accelerating the process of drug discovery from endolichenic fungi. In comparison with traditional methods these new techniques will provide rapid isolation, testing and use samples in less quantity for chemical profiling. Lately, it was noticed that there is a rising demand in the use of secondary metabolites isolated from endolichenic fungi for natural drug development however, nevertheless massive amount of struggle is essential to ascertain the capability of endolichenic fungi to produce bioactive metabolites. There is a lot of work yet to be done on the less documented inimitable pharmacological properties of endolichenic fungi. They are affluent mines of novel bioactive secondary metabolites and may personify the prospective biopharmaceuticals.

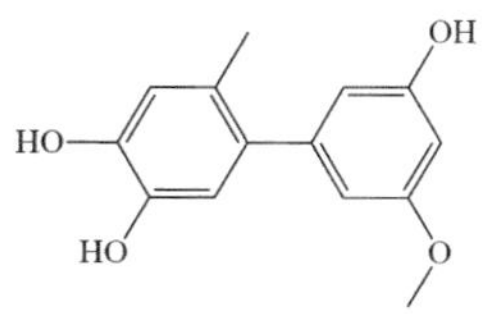
5'-Methoxy-6-methyl-biphenyl-3,4,3'-triol **1**

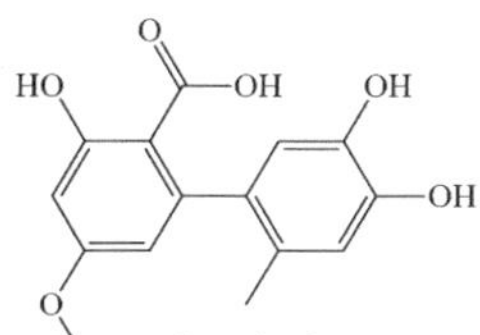
Altenusin **2**

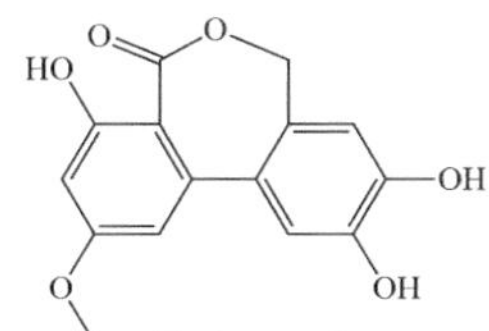
Alterlactone **3**

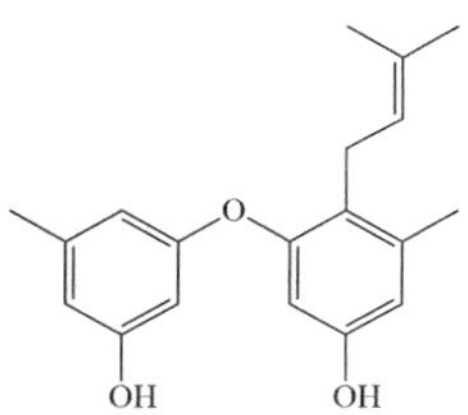
2-Isopentenyldiorcinol **4**

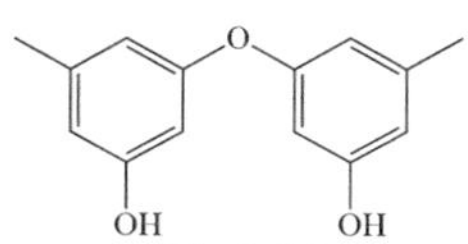
Diorcinol **5**

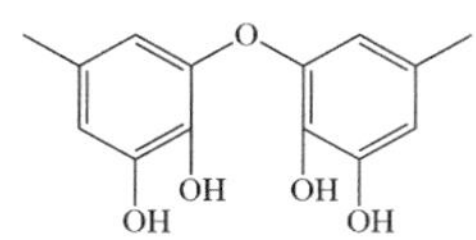
Violaceol-I **6**

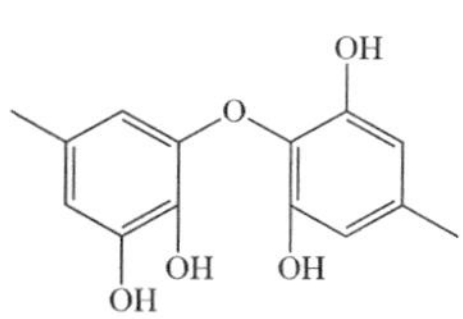
Violaceol-II **7**

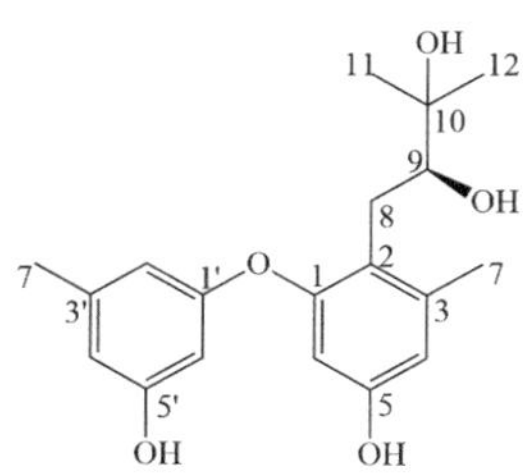
Diorcinol B **8**

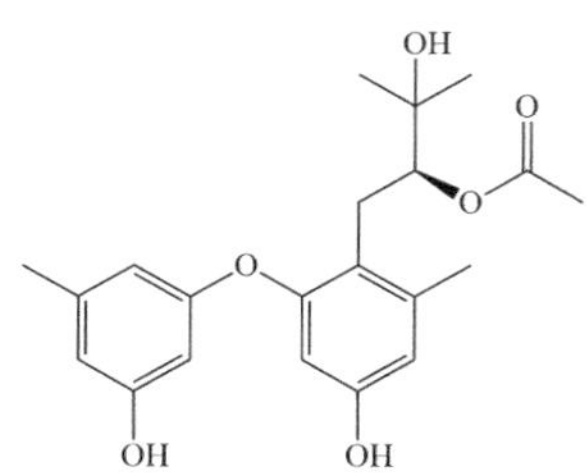
9-Acetyldiorcinol B **9**

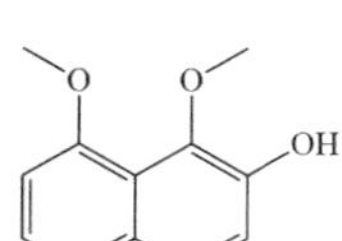
1,8-Dimethoxynaphthalen-2-ol **10**

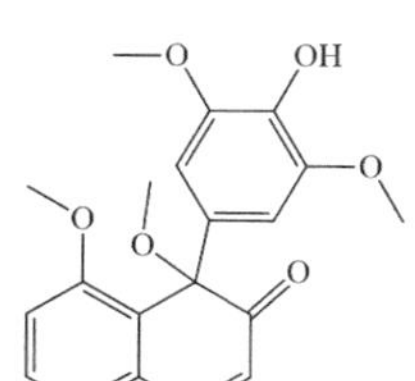
1-(4 -Hydroxy-3 ,5 -dimethoxy-phenyl)-1,8-dimethoxynaphthalen-2(1H)-one **11**

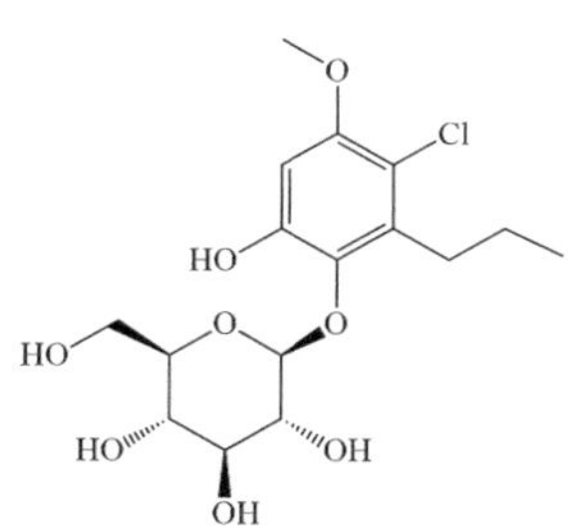
Globosumoside A **12**

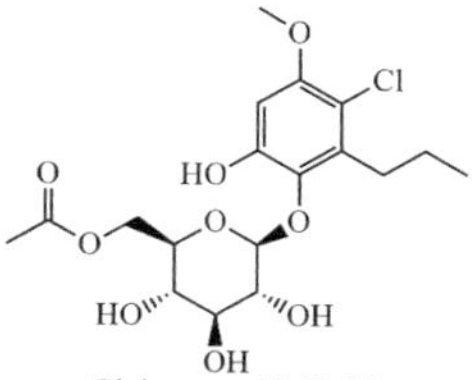

Globosumoside B **13**

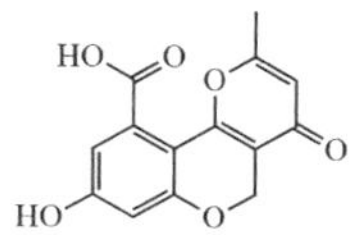

Myxotrichin D **14**

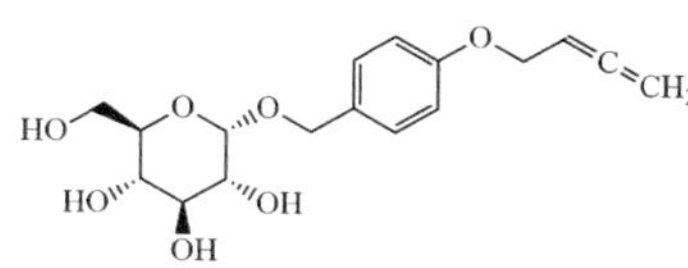

Terricollene A **15**

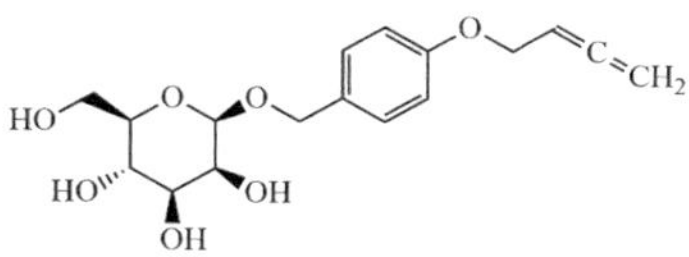

Terricollene B **16**

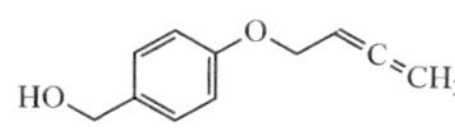

Terricollene C **17**

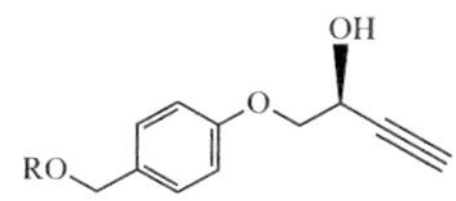

R = H Terricolyne **18**
R = CH_3 1-O-Methylterricolyne **19**
R = Ac 1-O-Acetylterricolyne **20**

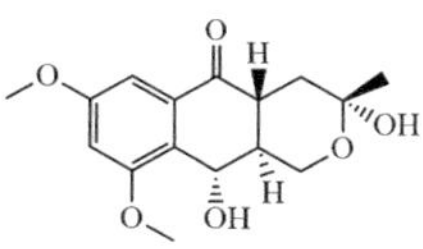

Corynesporol **21**

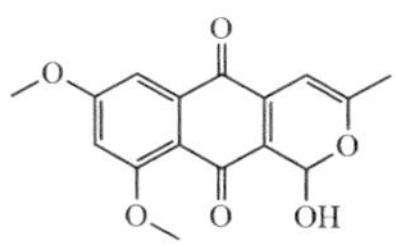

1-Hydroxydehydroherbarin **22**

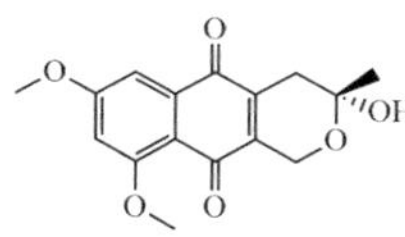

Herbarin **23**

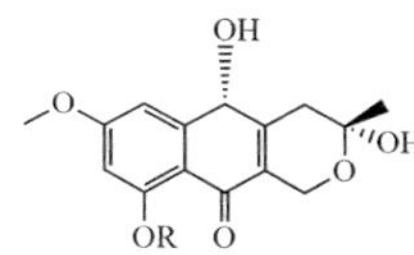

R = CH_3 9-O-Methylscytalol A **24**
R = H Scytalol A **27**

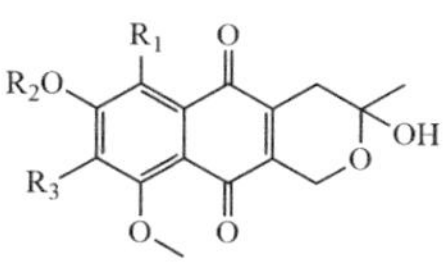

R_1= R_2= R_3= H 7-Desmethylherbarin **25**
R_1= H R_2= CH_3 R_3= OH 8-Hydroxyherbarin **26**
R_1= OH R_2= CH_3 R_3= H 8-O-Methylfusarubin **28**

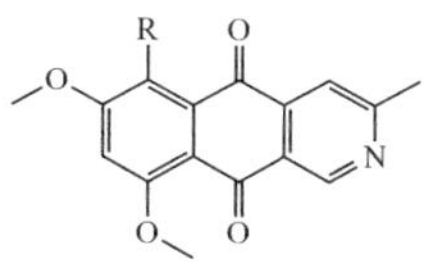

R = H Scorpinone **29**
R = OH 8-O-Methylbostrycoidin **30**

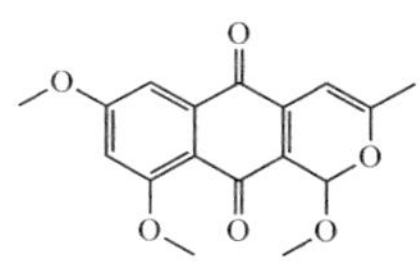

1-Methoxydehydroherbarin **31**

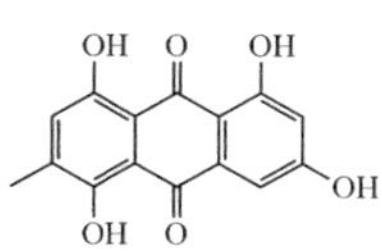

Catenarin **32**

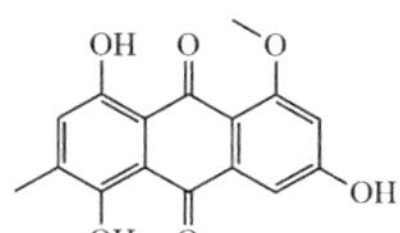

Rubrocristin **33**

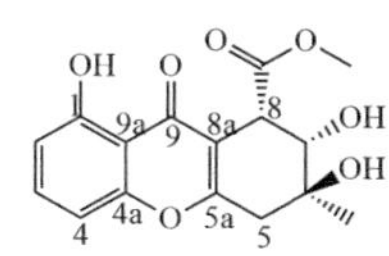

Preussochromone B **34**

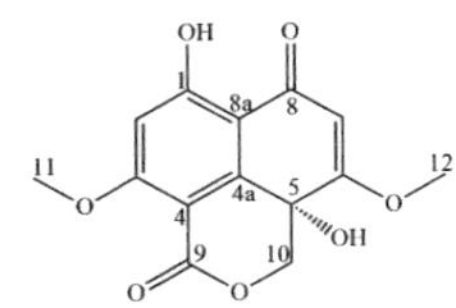
Preussochromone C **35**

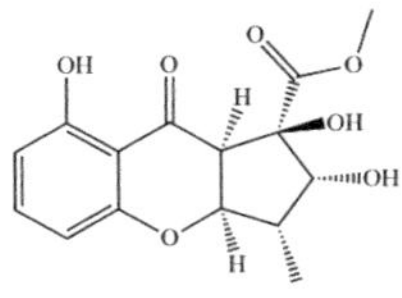
Preussochromone D **36**

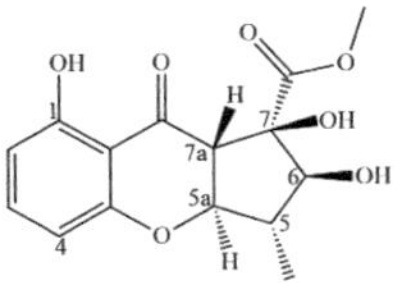
Preussochromone E **37**

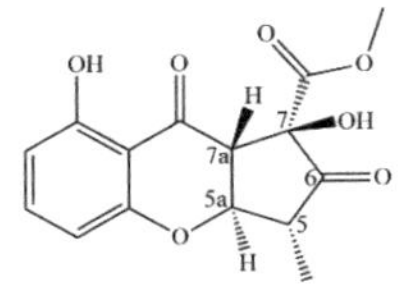
Preussochromone F **38**

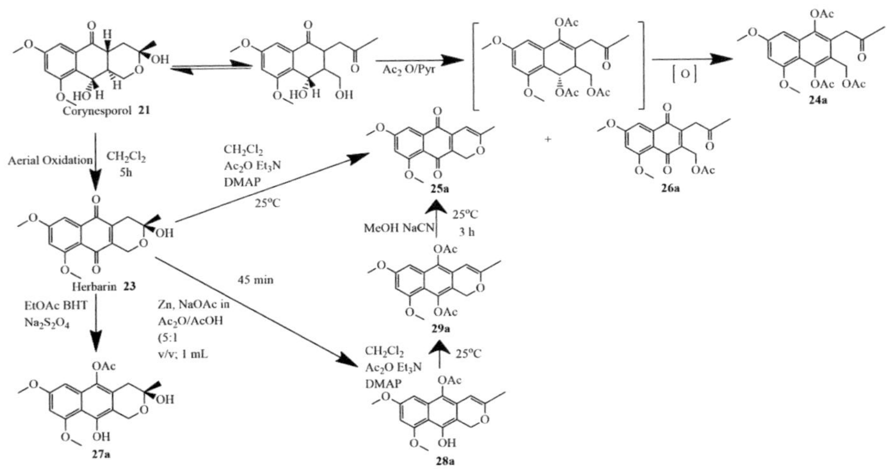

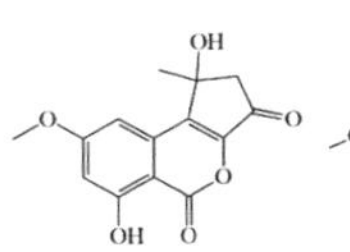
Rubralactone **39**

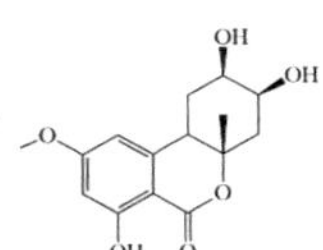
Dihydroaltenuene A **40**

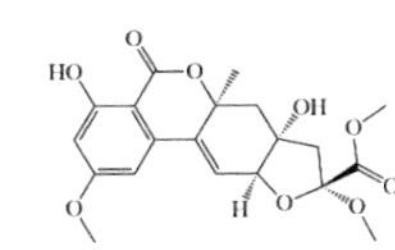
Xinshengin **41**

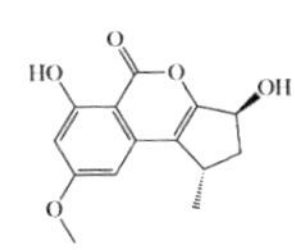
Phialophoriol **42**

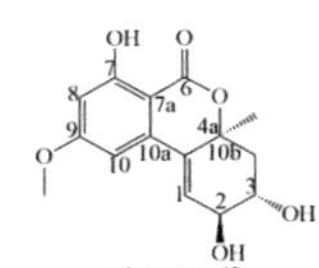
Altenuene **43**

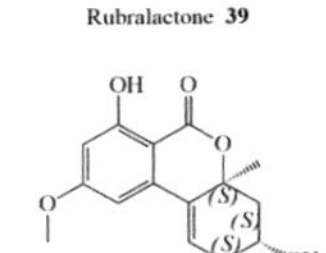
(+)-(2S,3S,4aS)-Altenuene **43a**

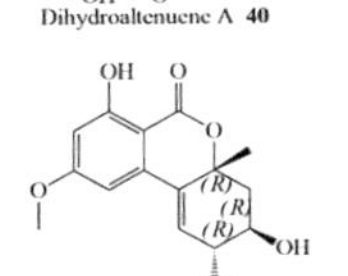
(-)-(2R,3R,4aR)-Altenuene **43b**

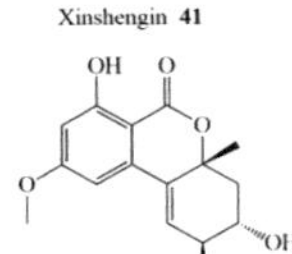
Isoaltenuene **44**

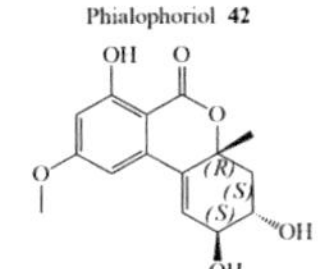
(-)-(2S,3S,4aR)-Isoaltenuene **44a**

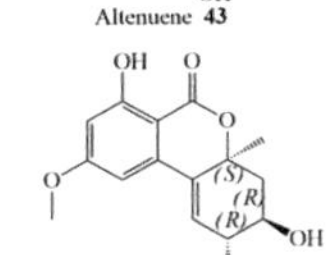
(+)-(2R,3R,4aS)-Isoaltenuene **44b**

$R_1 = R_2 = H$ Alternariol **45**
R_1 = Me R_2 = H Alternariol-9-methyl ether **46**
R_1 = Me R_2 = OH 4-Hydroxyalternariol-9-methyl ether **47**

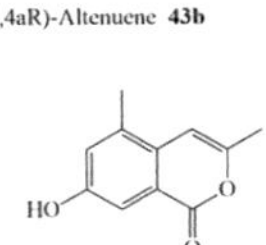
7-Hydroxy-3, 5-dimethyl-isochromen-1-one **48**

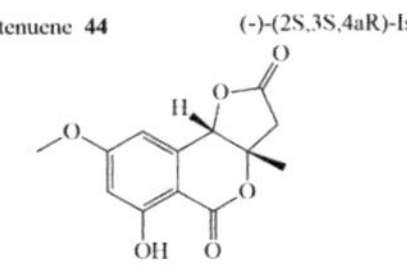
6-Hydroxy-8-methoxy-3a-methyl-3a,9b-dihydro-3H-furo[3,2-c]isochromene-2,5-dione **49**

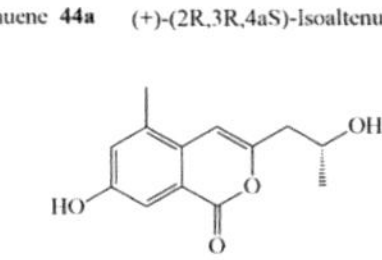
7-Hydroxy-3-(2-hydroxy-propyl)-5-methyl-isochromen-1-one **50**

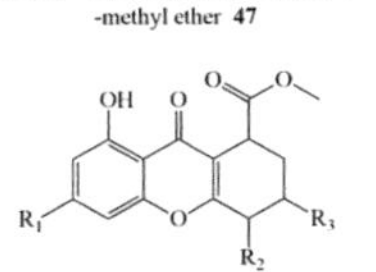
$R_1 = CH_2OH$ R_2 = H R_3 = OH Conioxanthone A **51**
$R_1 = CH_3$ R_2 = OH R_3 = H Microxanthone **52**

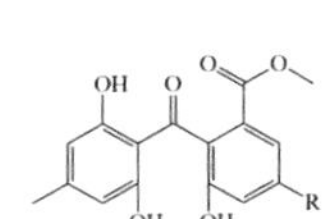
R = H Moniliphenone **53**
R = OCH_3 Isosulochrin **54**

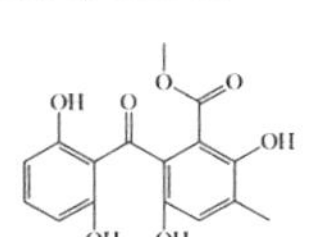
Sporormiellone A **55**

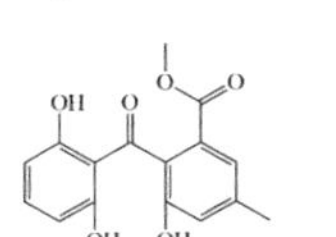
Sporormiellone B **56**

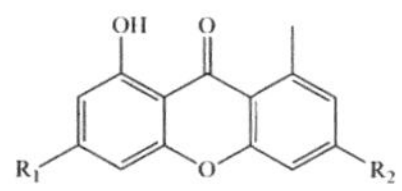

R_1 = OH R_2 = OH Norlichexanthone **57**
R_1 = OCH_3 R_2 = OH Griseoxanthone C **58**
R_1 = OH R_2 = OCH_3 6-O-Methylnorlichexanthone **59**

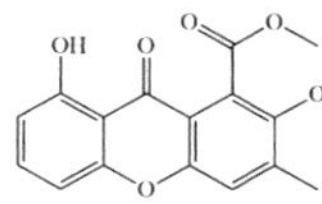

2,8-Dihydroxy-3-methyl-9-oxoxanthene-1-carboxylic acid methylester **60**

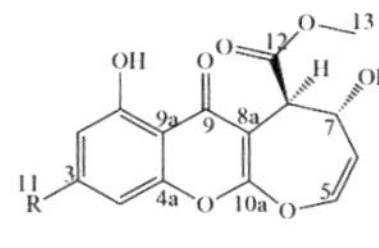

R = CH_3 Conioxepinol A **61**
R = CH_2OH Conioxepinol B **62**

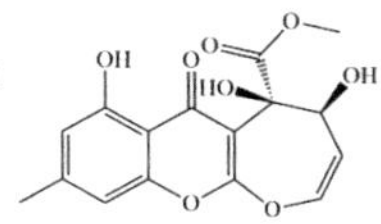

Conioxepinol C **63**

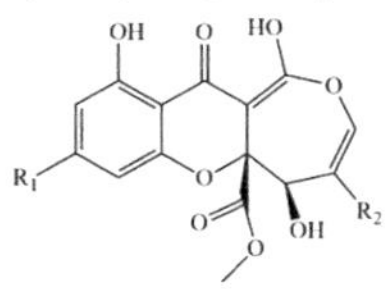

R_1 = CH_3 R_2 = H Conioxepinol D **64**
R_1 = H R_2 = CH_3 Brocaenol A **65**

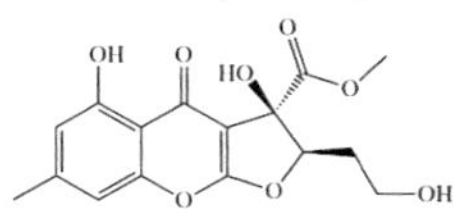

Coniofurol A **66**

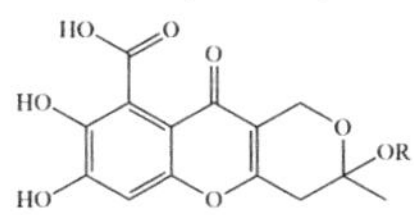

R = n-C_4H_9 Myxotrichin A **67**
R = CH_3 Myxotrichin B **68**

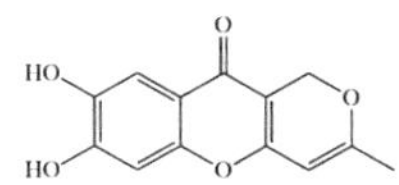

Myxotrichin C **69**

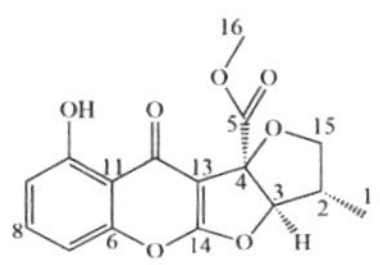

Sporormiellin A **70**

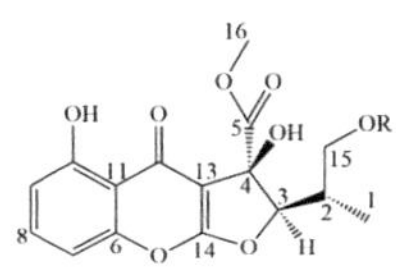

R = H Sporormiellin B **71**
R = $COCH_3$ Sporormiellin C **72**

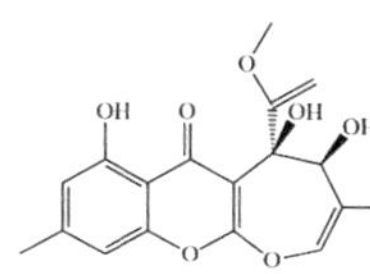

Microsphaeropsone A **73**

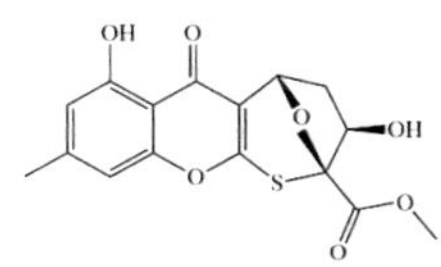

Coniothiepinol A **74**

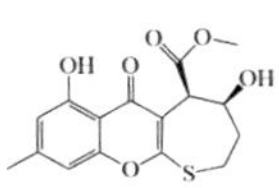

Coniothiepinol B **75**

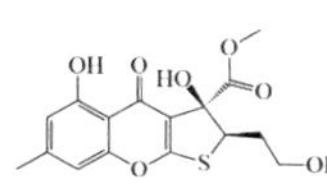

Coniothienol A **76**

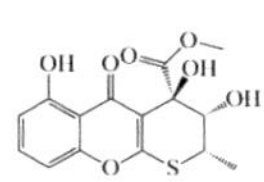

Preussochromone A **77**

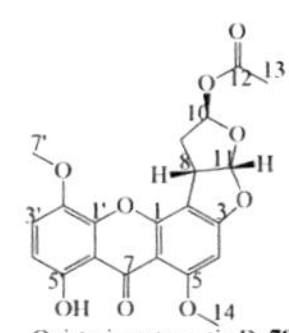

Oxisterigmatocystin D **78**

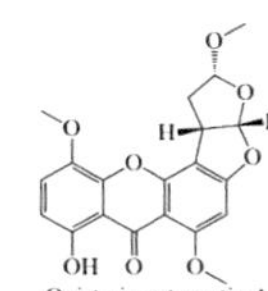

Oxisterigmatocystin A **79**

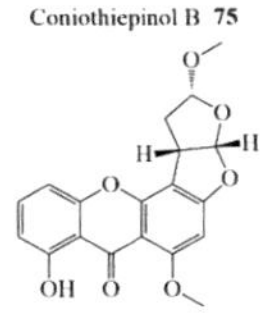

Oxisterigmatocystin C **80**

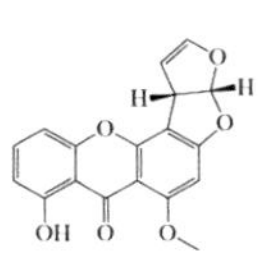

Sterigmatocystin **81**

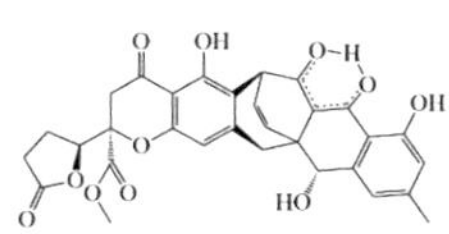

Xanthoquinodin A4 **82**

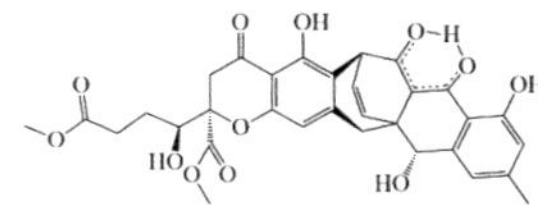

Xanthoquinodin A5 **83**

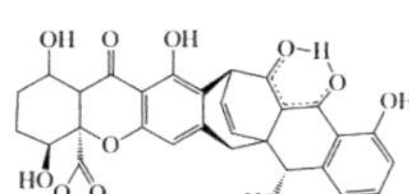

Xanthoquinodin A6 **84**

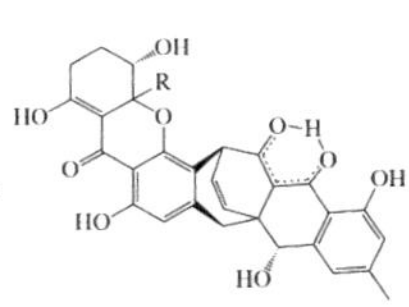

R = $COOCH_3$ Xanthoquinodin B4 **85**
R = $COOCH_3$ Xanthoquinodin B5 **86**

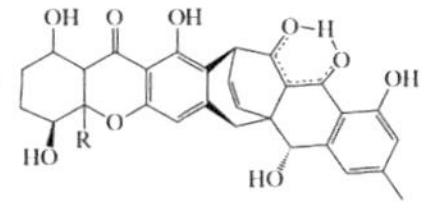

R = $COOCH_3$ Xanthoquinodin A1 **87**
R = $COOCH_3$ Xanthoquinodin A2 **88**

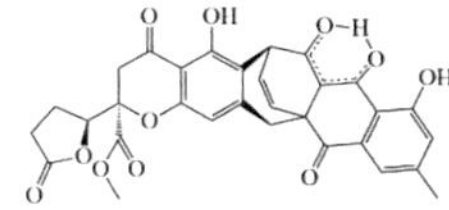

Xanthoquinodin A3 **89**

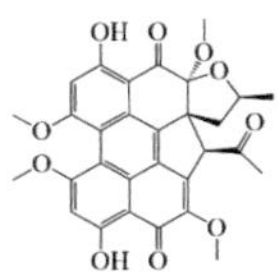
Phaeosphaerin A **90**

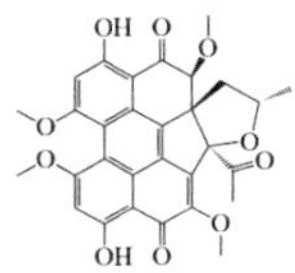
Phaeosphaerin B **91**

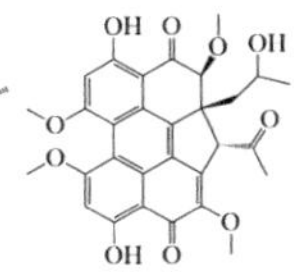
Phaeosphaerin C **92**

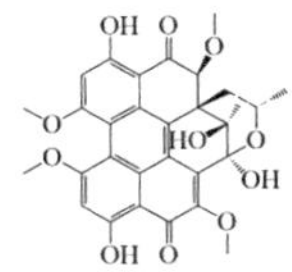
Phaeosphaerin D **93**

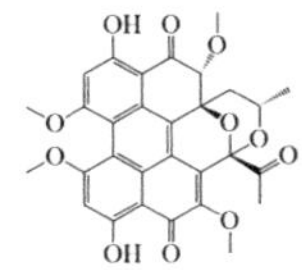
Phaeosphaerin E **94**

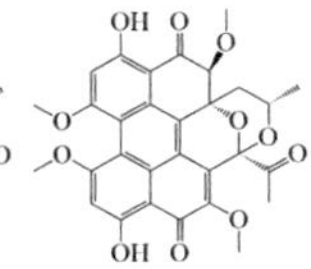
Phaeosphaerin F **95**

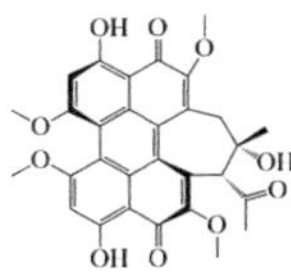
Hypocrellin A **96**

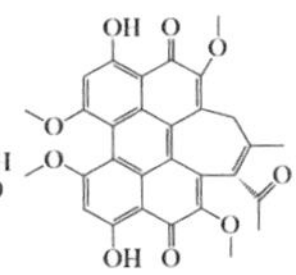
Hypocrellin C **97**

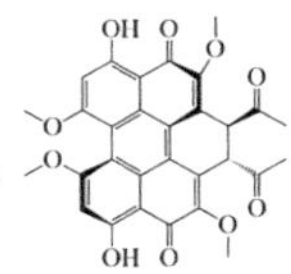
Elsinochrome A **98**

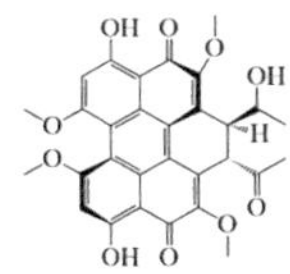
Elsinochrome B **99**

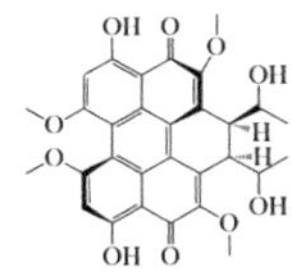
Elsinochrome C **100**

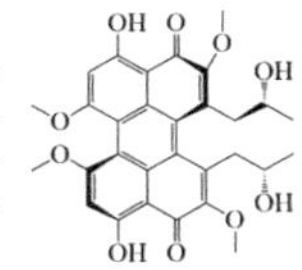
(+)-Calphostin D **101**

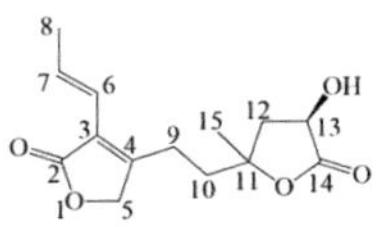
11*S* Pezizolide A **102**
11*R* Pezizolide B **103**

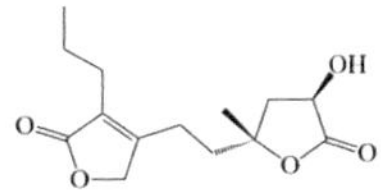
Pezizolide C **104**

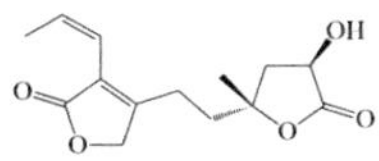
Pezizolide D **105**

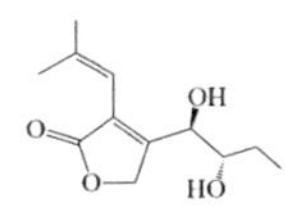
Pezizolide E **106**

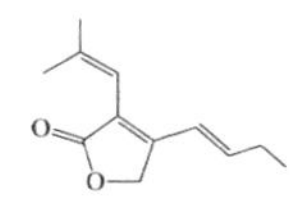
Pezizolide F **107**

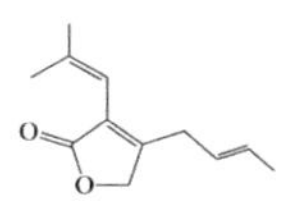
Pezizolide G **108**

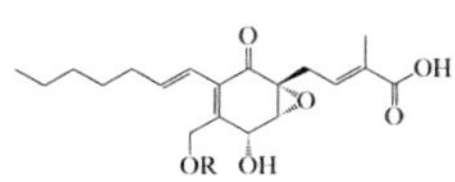
R = H Ambuic acid **109**
R = Ac **110**

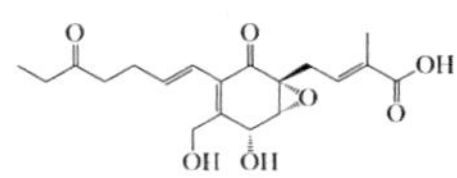
111

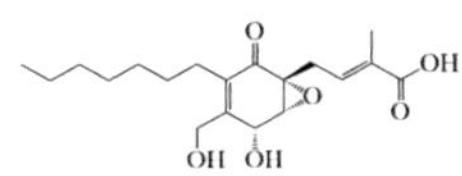
112

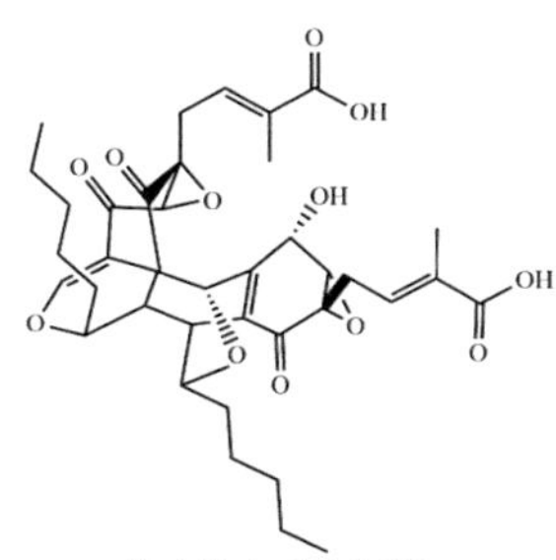
7-epi-Chaetoviridin E **120**

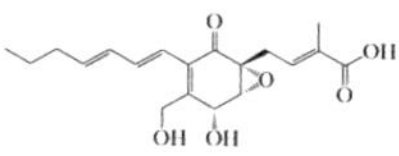

113

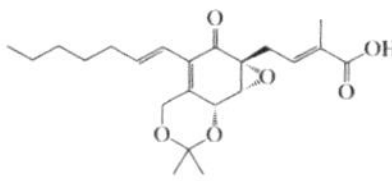

114

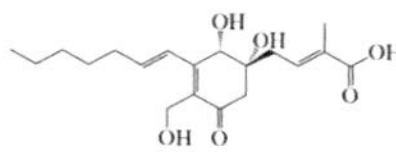

115

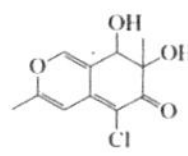

Myxodiol A **116**

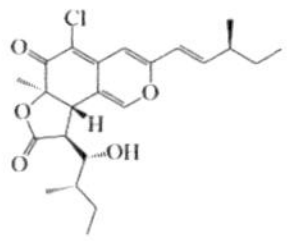

Chaetomugilin S **117**

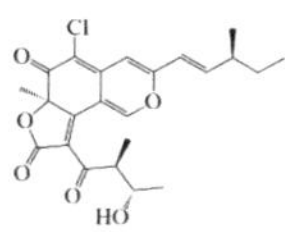

7,5'-bis-*epi*-Chaetoviridin A **118**

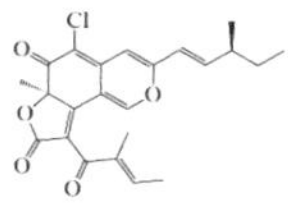

7-*epi*-Chaetoviridin E **119**

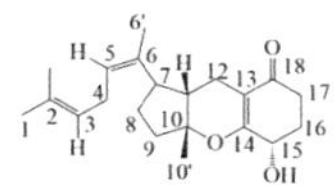

TCA F **121**

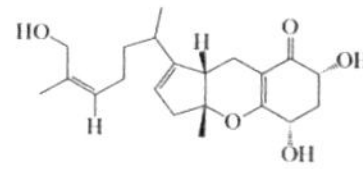

TCA G **122**

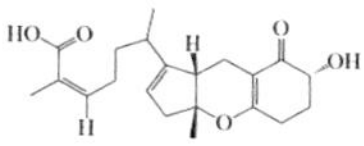

TCA H **123**

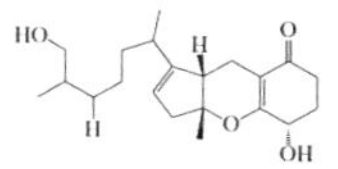

TCA 1a **124**

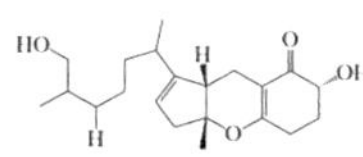

TCA 1b **125**

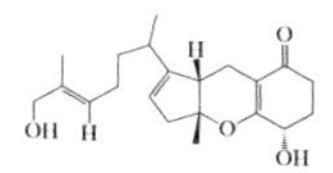

TCA 2a **126**

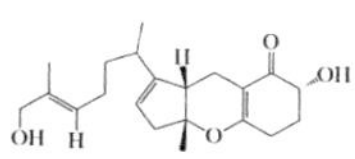

TCA 2b **127**

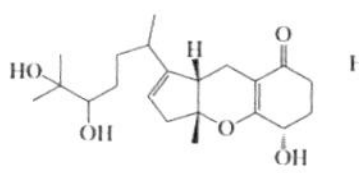

TCA 6a **128**

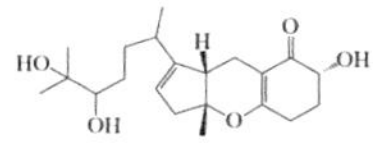

TCA 6b **129**

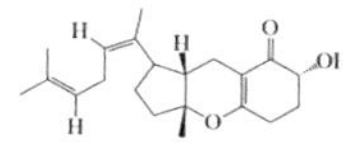

TCA 9b **130**

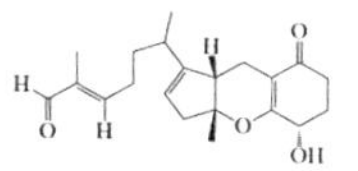

Tricycloalternarenal **131**

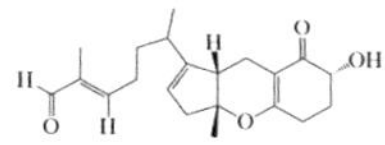

ACTG-toxin H **132**

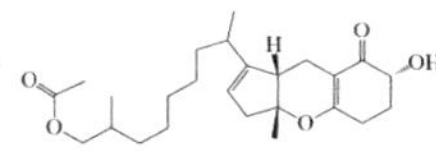

TCAD **133**

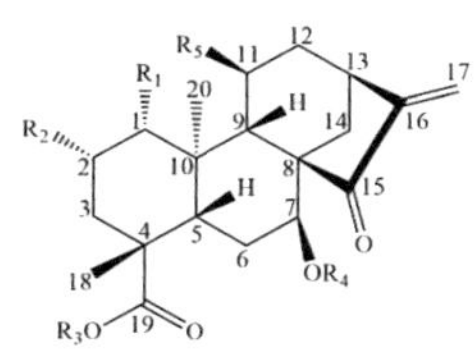

R_1 = OH R_2 = R3 = R_4 = R_5 = H Geopyxin A **134**
R_1 = R_2 = R_3 = R_4 = R_5 = H Geopyxin B **135**
R_1 = R_2 = R_3 = R_4 = H R_5 = OH Geopyxin D **137**
R_1 = OH R_2 = OH R3 = R_4 = R_5 = H Geopyxin E **138**

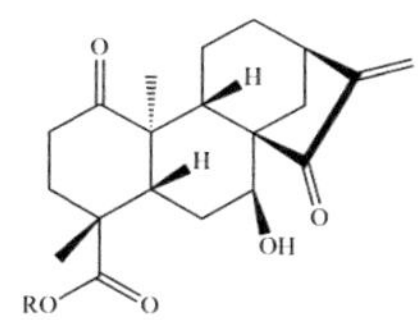

R = H Geopyxin C **136**

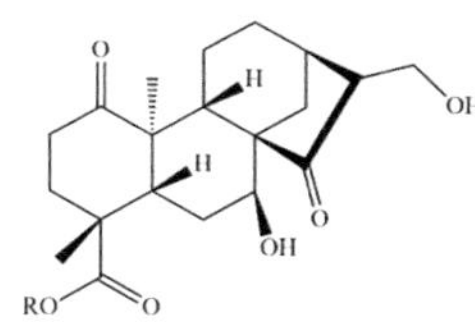

R = H Geopyxin F **139**

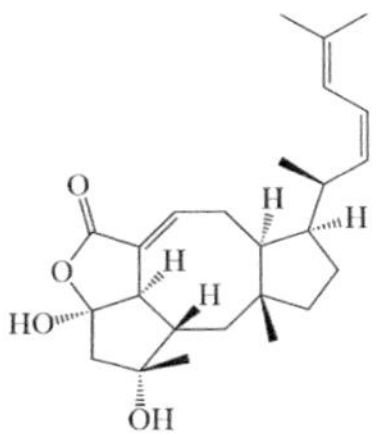

Ophiobolin P **140**

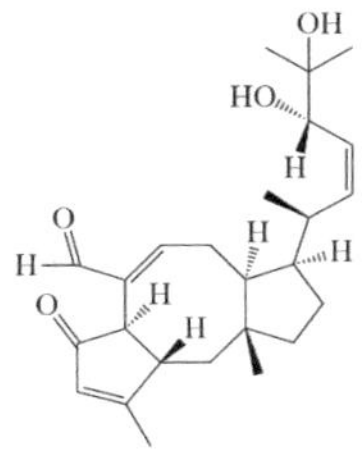

Ophiobolin Q **141**

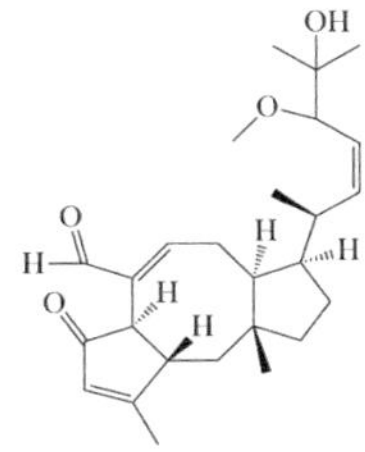

Ophiobolin R **142**

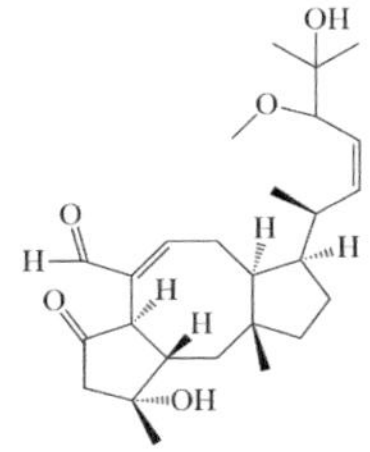

Ophiobolin S **143**

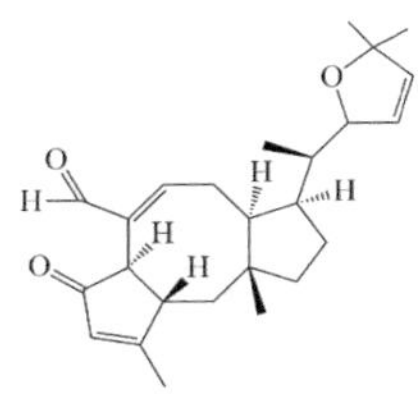

Ophiobolin T **144**

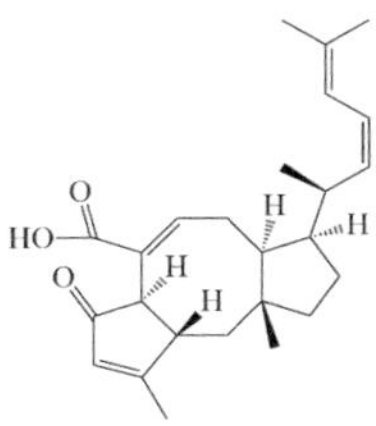

6-*epi*-21, 21-Odihydroophiobolin G **145**

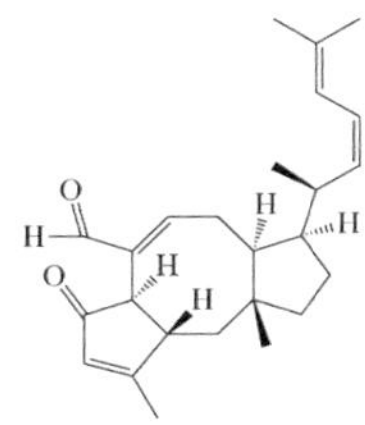

6-*epi*-Ophiobolin G **146**

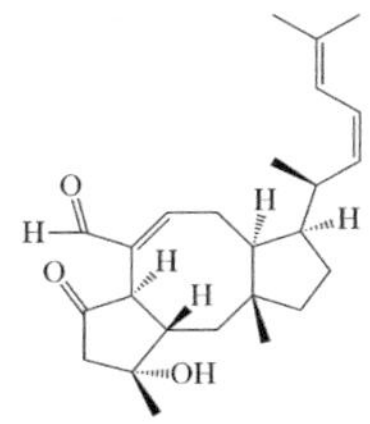

6-*epi*-Ophiobolin K **147**

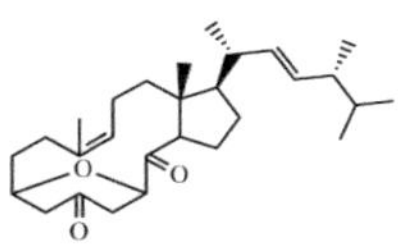

22E-3,7-Epoxy-5,10:8,9-disecoergosta-9(10), 22-diene-5,8-dione **148**

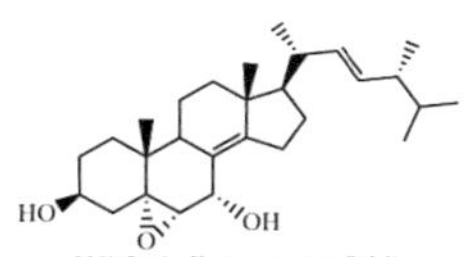

22E-5a,6a-Epoxyergosta-8(14), 22-diene-3ß,7a-diol **149**

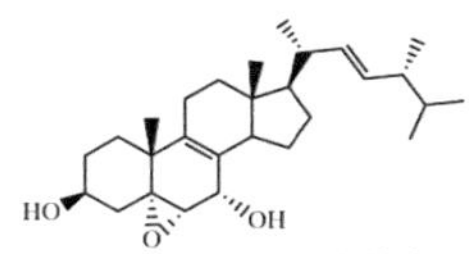

22E-5a,6a-Epoxyergosta-8(9),22-diene -3ß,7a-diol **150**

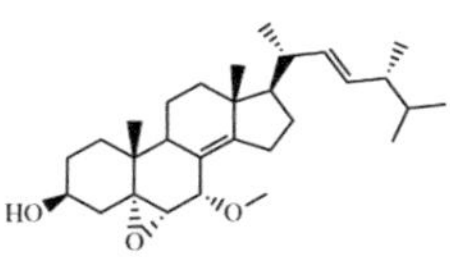

22E-7a-Methoxy-5a,6a-epoxyergosta-8(14), 22-dien-3ß-ol **151**

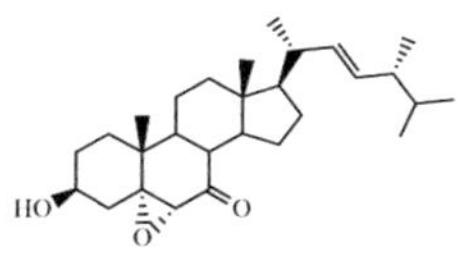

22E-3ß-Hydroxy-5a, 6a-epoxyergosta-22-en-7-one **152**

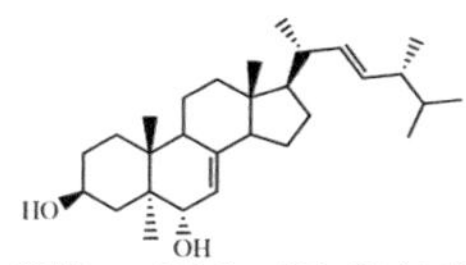

22E-Ergosta-7,22-diene-3ß,5a,6ß-triol **153**

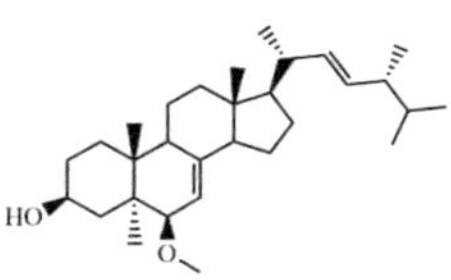

22E-6ß-Methoxyergosta-7, 22-diene-3ß,5a-diol **154**

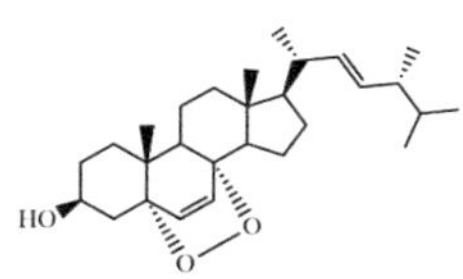

22E-5a,8a-Epidioxyergosta-6, 22-dien-3ß-ol **155**

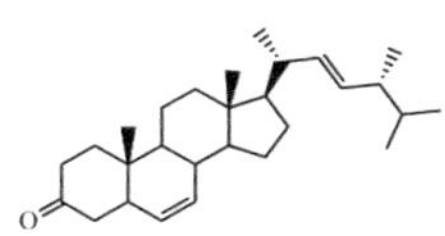

22E-Ergosta-4,6,8(14), 22-tetraen-3-one **156**

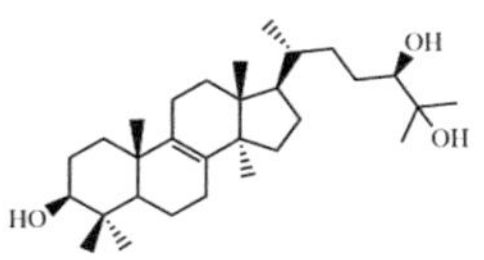

Inoterpene B **157**

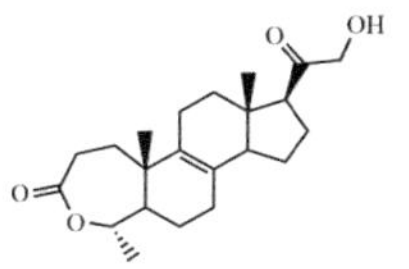

Nodulisporisteriod A **158**

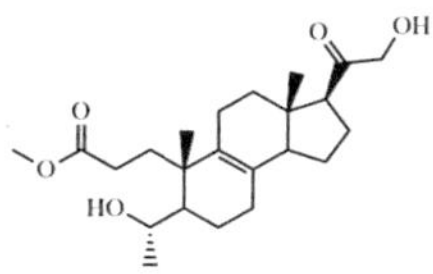

Nodulisporisteriod B **159**

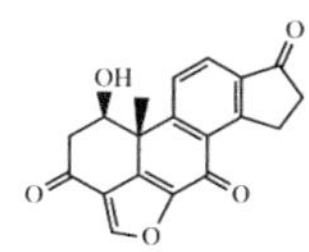

Demethoxyviridin **160**

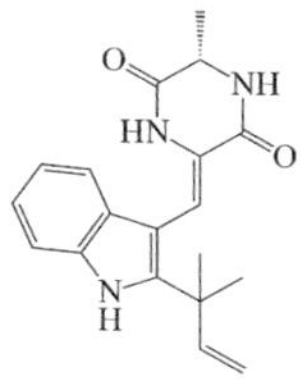

Neoechinulin A **161**

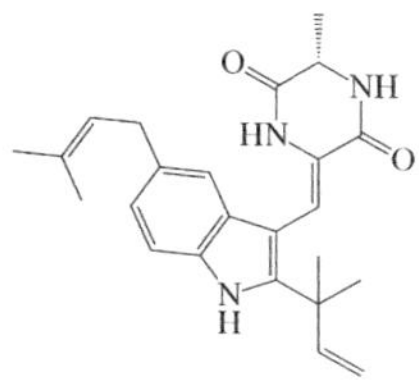

Isoechinulin A **162**

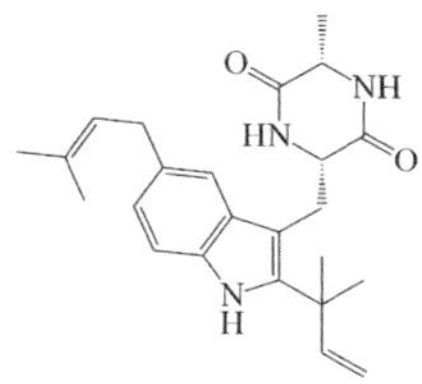

Tardioxopiperazine A **163**

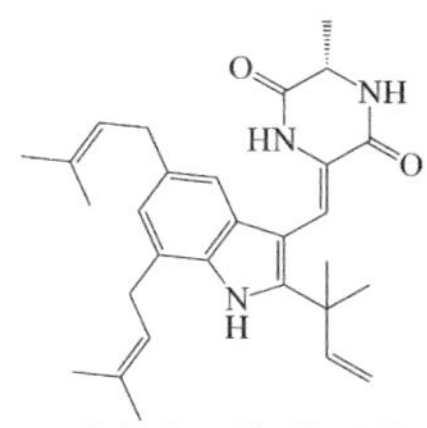

Dehydroechinulin **164**

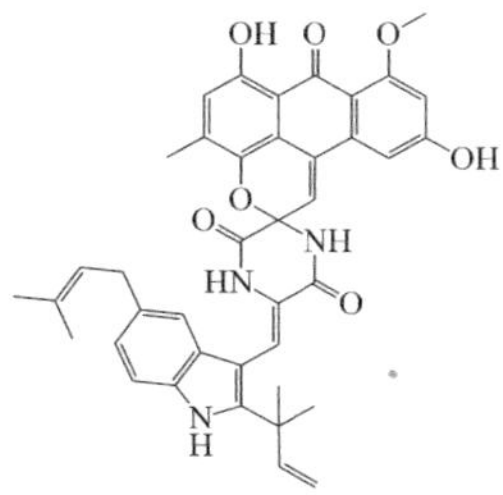

7-*O*-Methylvariecolortide A **165**

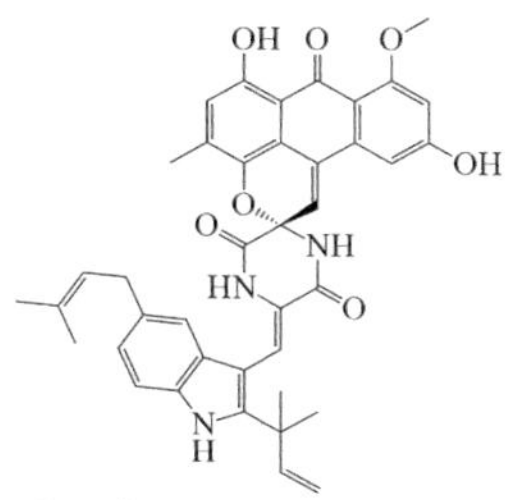

(-)-(*S*)-7-*O*-Methylvariecolortide A **165a**

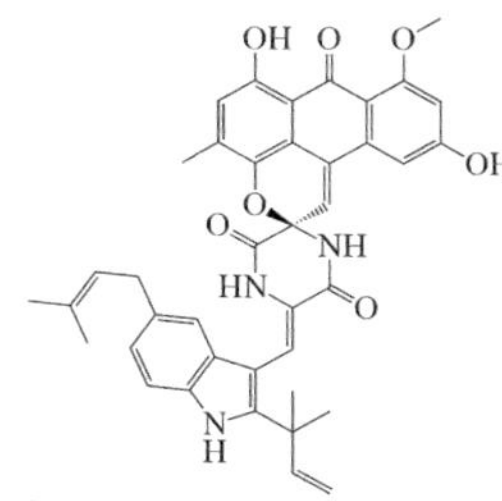

(+)-(*R*)-7-*O*-Methylvariecolortide A **165b**

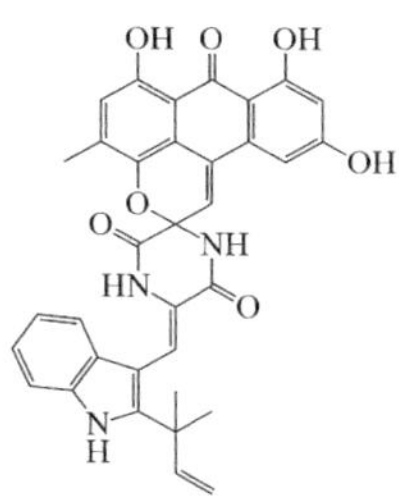

Variecolortide B **166**

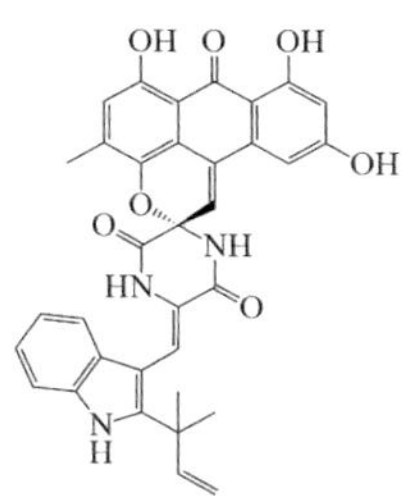

(-)-(*S*)-Variecolortide B **166a**

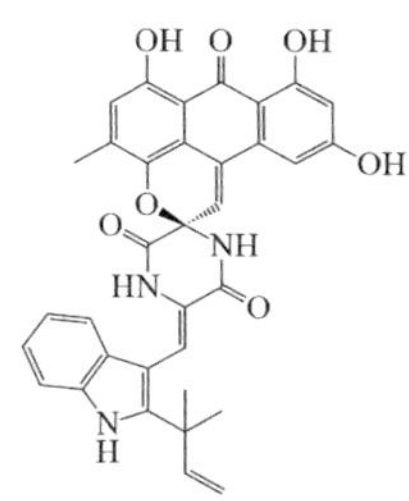

(+)-(*R*)-Variecolortide B **166b**

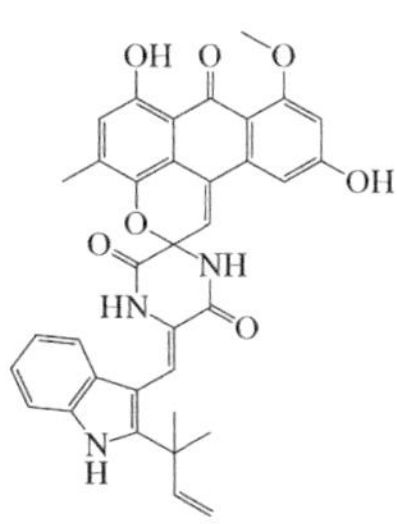

Variecolortide C **167**

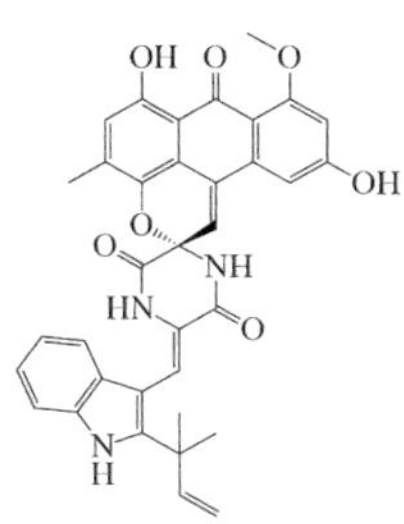

(-)-(S)-Variecolortide C **167a**

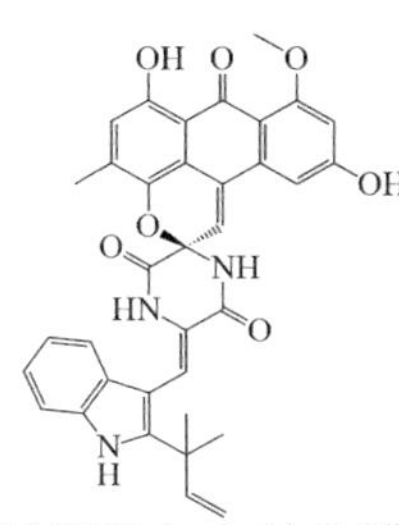

(+)-(R)-Variecolortide C **167b**

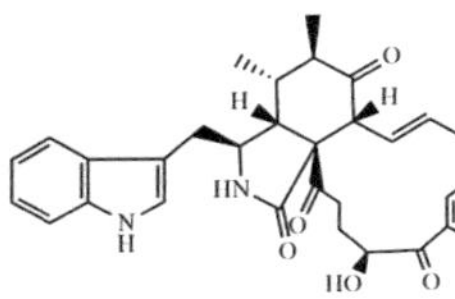

Chaetoglobosin Y **168**

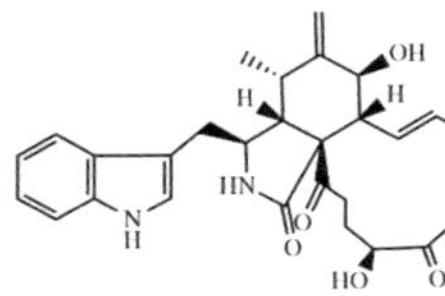

Chaetoglobosin F **169**

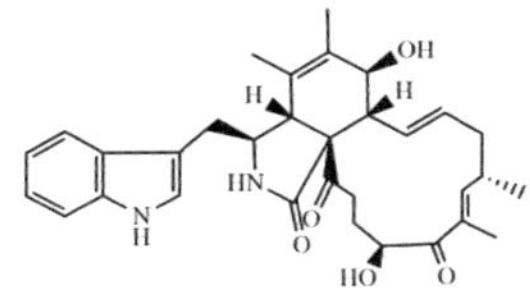

Chaetoglobosin E **170**

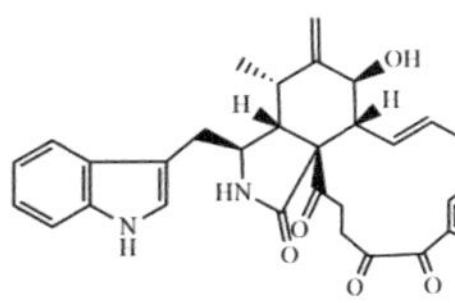

Chaetoglobosin D **171**

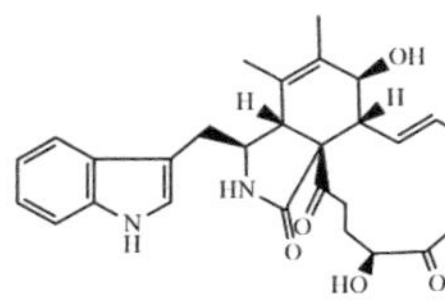

Chaetoglobosin G **172**

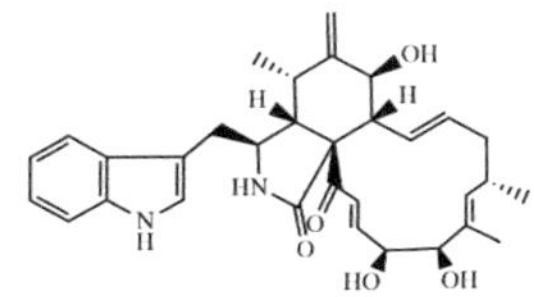

Cytoglobosin B **173**

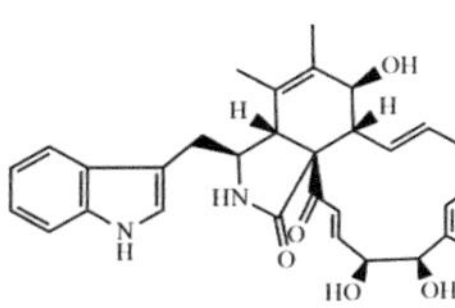

Cytoglobosin C **174**

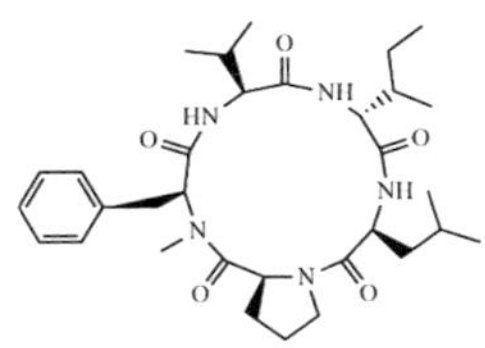

Cyclo(N-methyl-L-Phe-L-Val-D-Ile-L-Leu-L-Pro) **175**

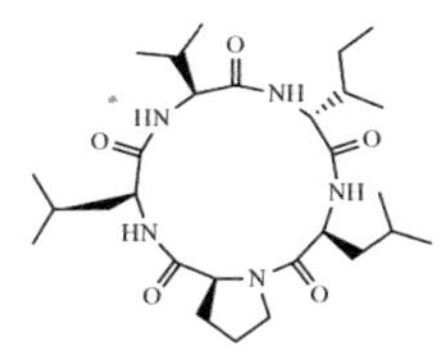

Cyclo(L-Val-D-Ile-L-Leu-L-pro-D-Leu) **176**

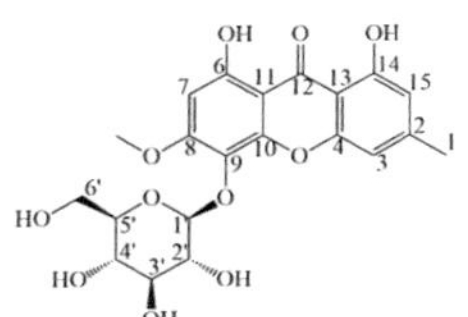

Sporormielloside **177**

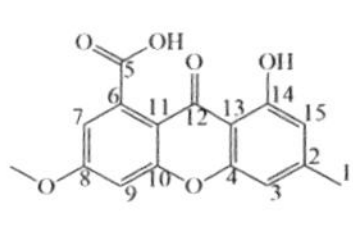

Calyxanthone-8-methyl ether **178**

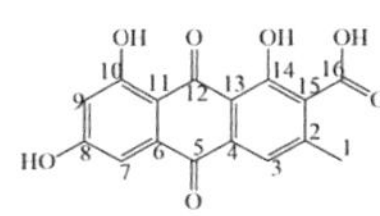

Endocrocin **179**

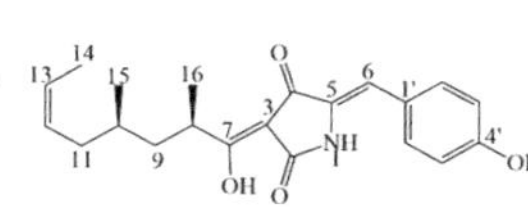

Tolypocladenols A1 **180**

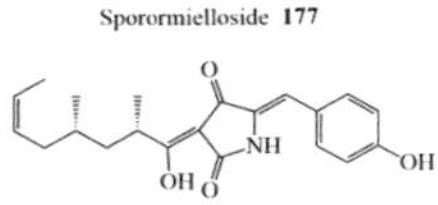

Tolypocladenols A2 **181**

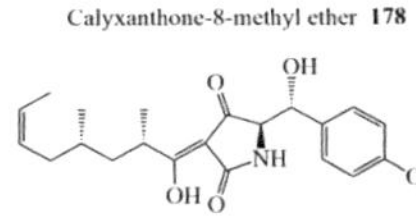

Tolypocladenols B **182**

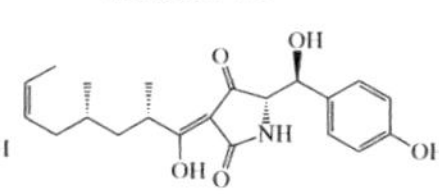

F-14329 **185**

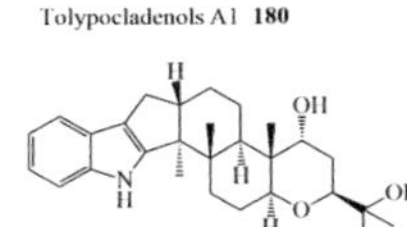

Terpendole E **188**

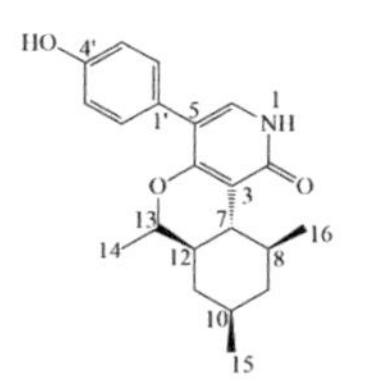

Tolypyridone A **183**

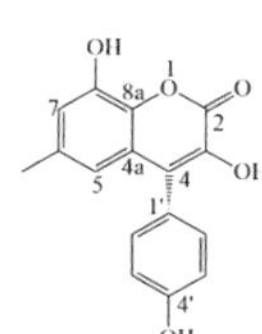

3,8-dihydroxy-4-(4-hydroxyphenyl)
-6-methylcoumarin **184**

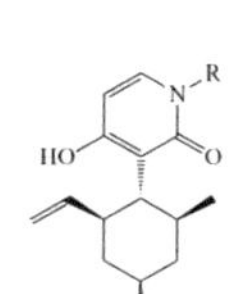

R = OH Pyridoxatin **186**
R = H Tolypyridone B **187**

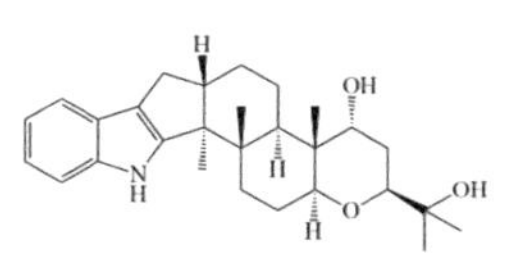

Terpendole E **188**

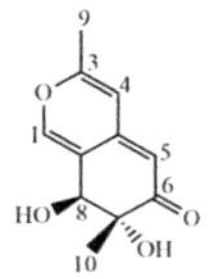

7,8-dihydro-7R,8S-dihydroxy-3,7-dimethyl-2-benzopyran-6-one **189**

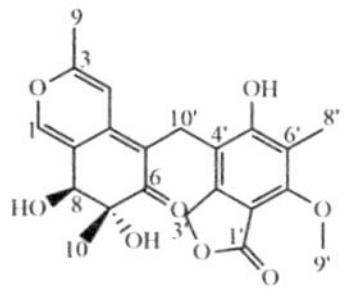

Myxotritones A **190**

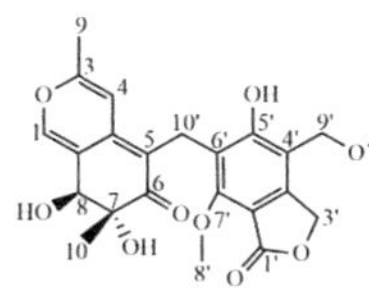

Myxotritones B **191**

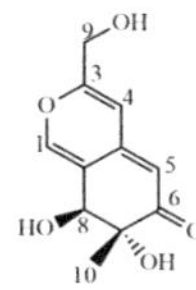

Myxotritones C **192**

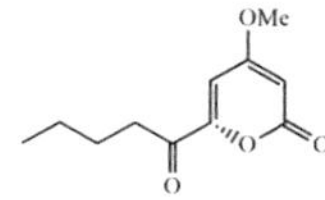

(*S*)-4-methoxy-6-pentanoyl-5,6-dihydro-2H-pyran-2-one **202**

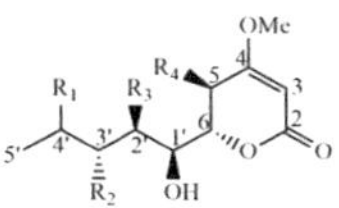

R_1 = OH (4′*S*) R_2 = R_3 = R_4 = H Necpyrones A **193**
R_1 = OH (4′*R*) R_2 = R_3 = R_4 = H Necpyrones B **194**
R_1 = R_3 = R_4 = H R_2 = OH Necpyrones D **196**
R_1 = R_2 = R_3 = H R_4 = OH Necpyrones E **197**
R_1 = R_2 = R_3 = R_4 = H (6S, 1′S)-LL-P880a **199**
R_1 = R_2 = R_4 = H R_3 = OH (6S, 1′*S*, 2′*R*)-LL-P880b **200**
R_1 = R_2 = R_4 = H R_3 = OAc (1′S, 2′R)-1-hydroxy-1-((*S*)-4-methoxy-6-oxo-3,6-dihydro-2H-pyran-2-yl)-pentan-2-yl acetate **201**

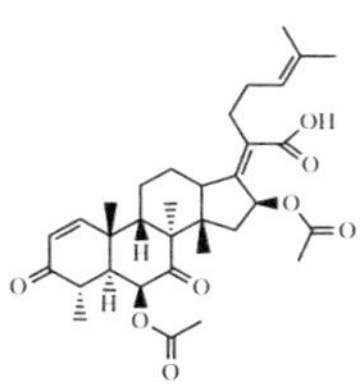

Helvolic acid **204**

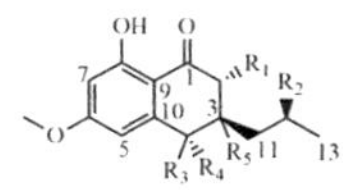

R_1 = CH_3, R_3 = R_5 = H, R_2 = R_4 = OH Biatriosporin A **207**
R_1 = R_3 = R_5 = H, R_2 = R_4 = OH Biatriosporin B **208**
R_1 = R_4 = R_5 = H, R_2 = R_3 = OH Biatriosporin C **209**
R_1 = R_3 = R_5 = H, R_2 = O, R_4 = OH **220**
R_1 = R_4 = R_5 = H, R_2 = O, R_4 = OH **221**
R_1 = R_4 = H, R_2 = R_3 = R_5 = OH **222**

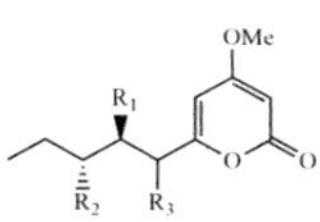

R_1 = H R_2 = OH R_3 = OH (1′R) Necpyrones C **195**
R_1 = R_2 = H R_3 = OH (1′R) PC-2 **198**
R_1 = OH R_2 = H R_3 = OH (1′*S*) (1′*S*, 2′*R*)-LL-P880*γ* **203**

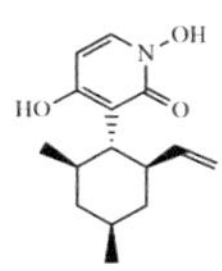

Pyridoxatin **205**

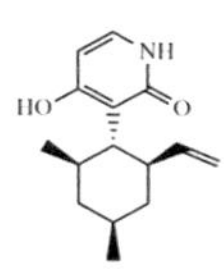

N-deoxy-pyridoxatin **206**

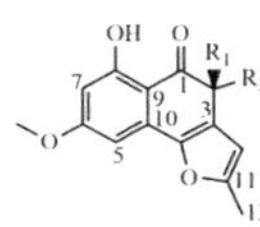

R_1 = R_2 = O Biatriosporin D **210**
R_1 = OH, R_2 = CH_3 Biatriosporin E **211**
R_1 = OH, R_2 = CH_2OCH_3 Biatriosporin F **212**

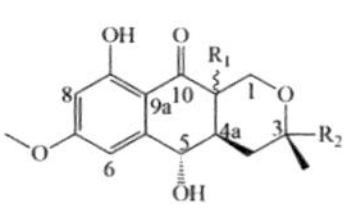

R_1 = β-H, R_2 = OCH_3 Biatriosporin G **213**
R_1 = α-H, R_2 = H Biatriosporin I **215**

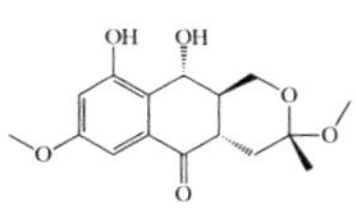

Biatriosporin H **214**

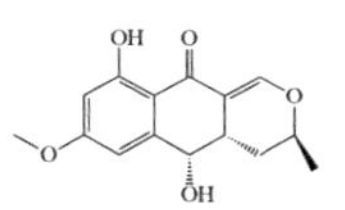

Biatriosporin J **216**

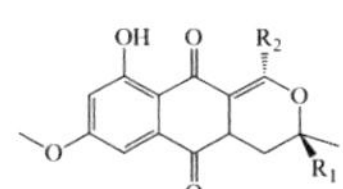

R_1 = α-H, R_2 = OCH_3 Biatriosporin K **217**
R_1 = OH, R_2 = H 6-O-demethyl-5-deoxyfusarubin **223**
R_1 = α-H, R_2 = H Pyranonaphthoquinone **224**

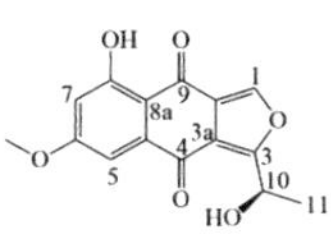

Biatriosporin L **218**

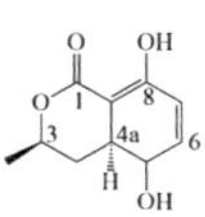

Biatriosporin M **219**

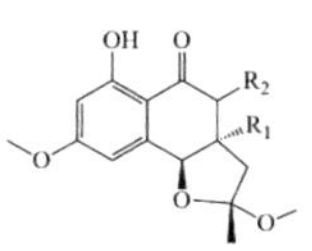

R_1 = OH, R_2 = H Balticols F **225**
R_1 = H, R_2 = Scytalols **226**

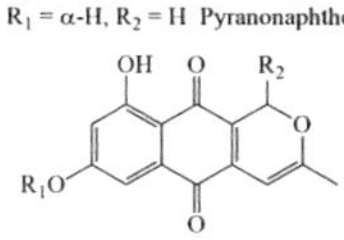

R_1 = H, R_2 = H 6-deoxy-7-O-demethyl-3,4-anhydrofusarubin **227**
R_1 = CH_3, R_2 = H 6-deoxy-3,4-anhydrofusarubin **228**
R_1 = CH_3, R_2 = OCH_3 ascomycone A **229**

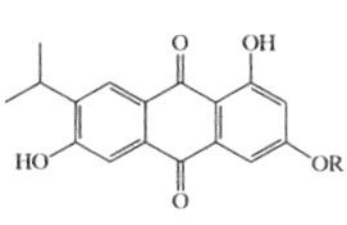

R = H ZSU-H85 A **230**
R = CH_3 3-acetyl-2,8-dihydroxy-6-methoxy anthraquinone **231**

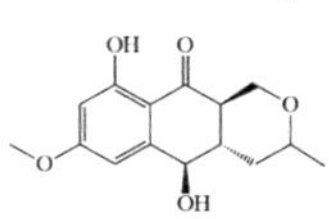

Pleorubrin B **232**

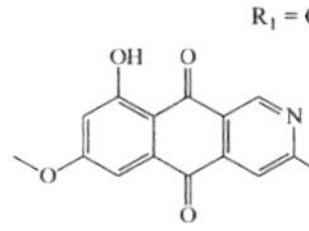

6-deoxybostrycoidin **233**

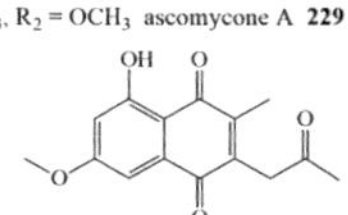

2-acetonyl-3-methyl-5-hydroxy-7-methoxynaphthazarin **234**

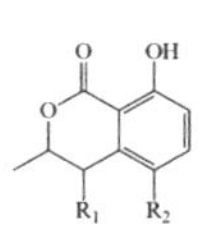

R_1 = H, R_2 = OH 5-hydroxymellein **235**
R_1 = OH, R_2 = H 4-hydroxymellein **236**
R_1 = OH, R_2 = OH 3,4-dihydro-4,5,8-trihydroxy-3-methylisocoumarin **237**

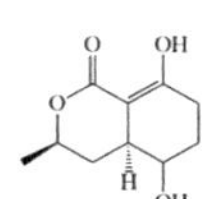

5-hydroxyramulosin **238**

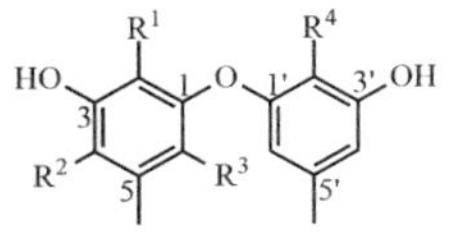

	R^1	R^2	R^3	R^4
239	MeO	H	H	OH
240	prenyl	prenyl	H	H
253	prenyl	H	H	H
254	H	H	H	H
255	H	H	prenyl	H
256	OH	H	H	OH
257	OH	H	H	H

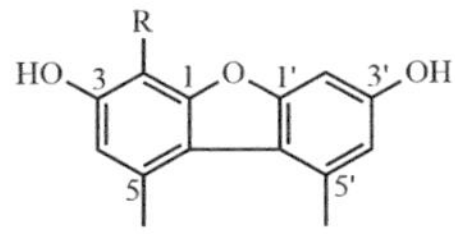

241 R = MeO
258 R = H

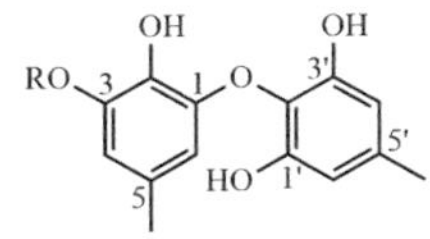

242 R = Me
259 R = H

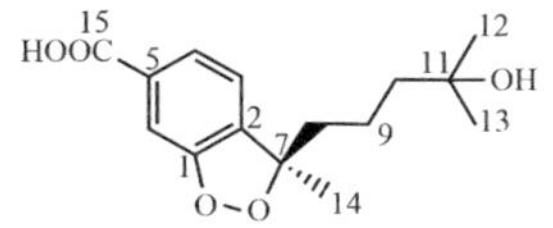

243

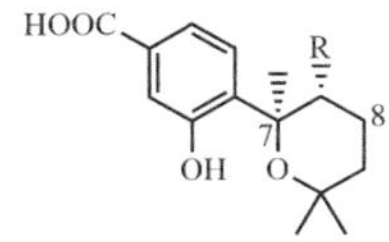

R = OH **244**
R = H **260**

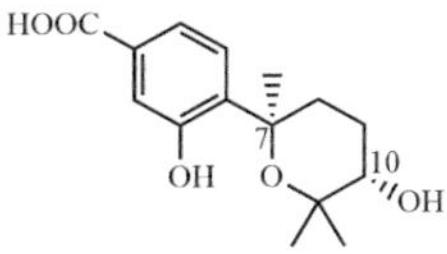

(-)-(7R,10S)-10-hydroxysydowic acid **245**

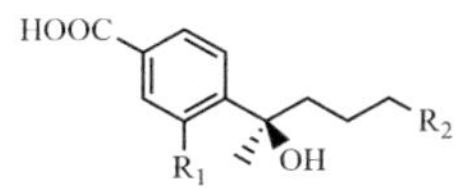

247 R^1 = H, R^2 =

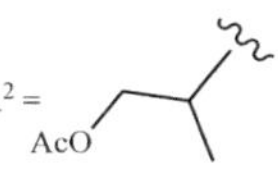

248 R^1 = OH, R^2 =

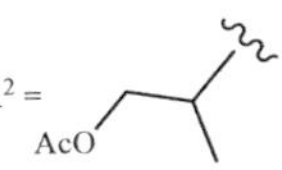

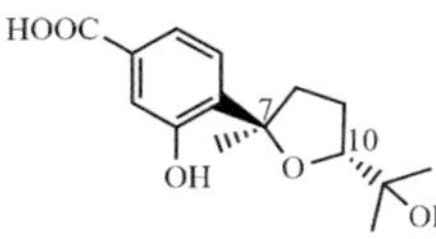

(-)-(7R,10R)-iso-10-hydroxysydowic acid **246**

249 R^1 = OH, R^2 =

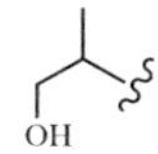

250 R^1 = OH, R^2 =

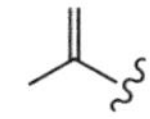

261 R^1 = OH, R^2 =

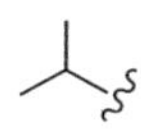

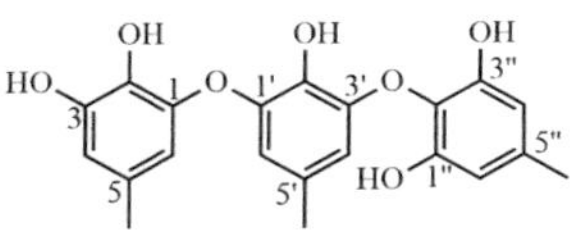

Sydowiol D **251**

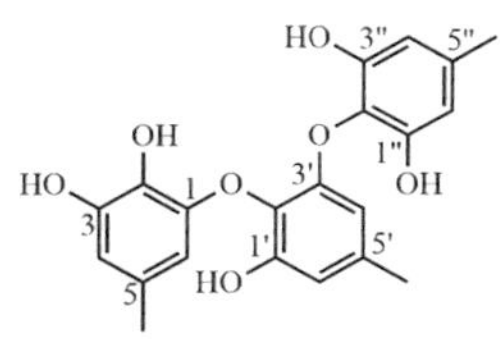

Sydowiol E **252**

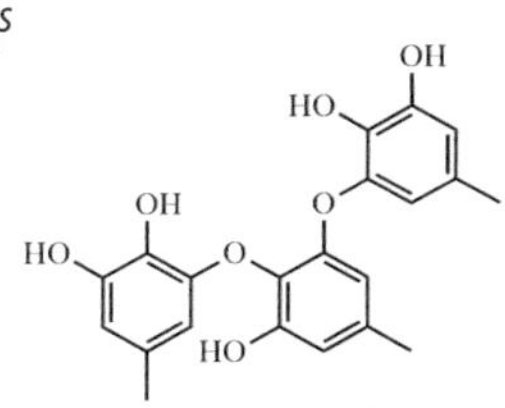

Sydowiol B **263**

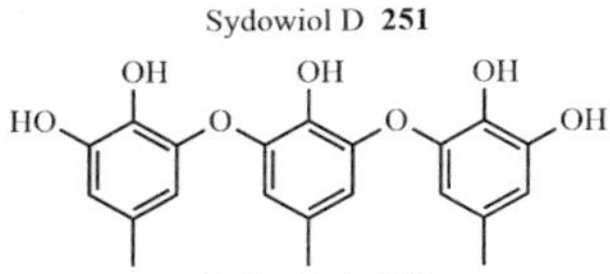

Sydowiol A **262**

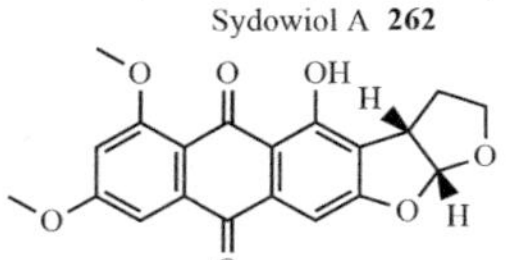

Aversin **264**

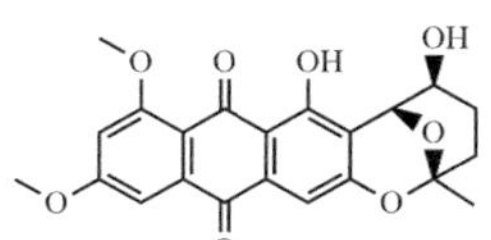

6,8-di-*O*-methylnidurufin **265**

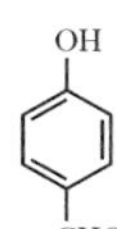

4-hydroxybenzaldehyde **266**

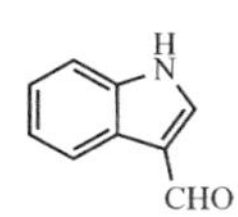

3-formyl-1H-indole **267**

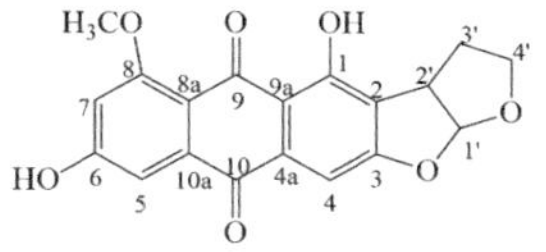

8-*O*-methylversicolorin B **268**

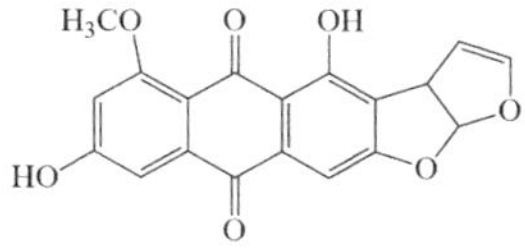

8-*O*-methylversicolorin A **269**

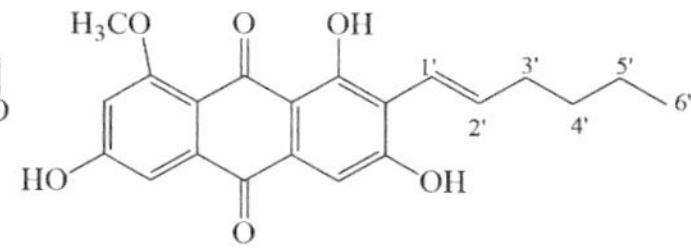

8-*O*-methylaverythin **270**

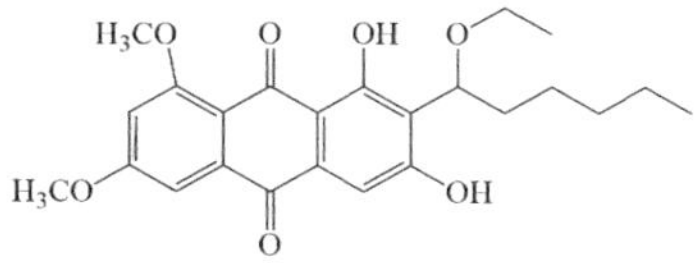

1'-*O*-ethyl-6,8-di-*O*-methylaverantin **271**

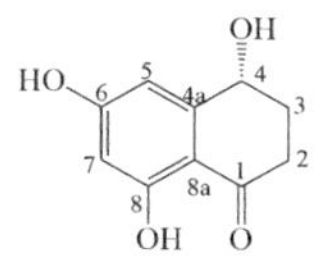

(*R*)-4,6,8-trihydroxy-3,4-dihydro-1 (2H)-naphthalenone **272**

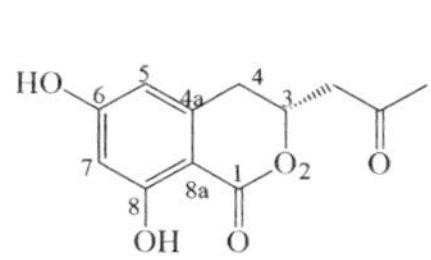

6,8-dihydroxy-(3R)-(2-oxopropyl)-3,4-dihydroisocoumarin **273**

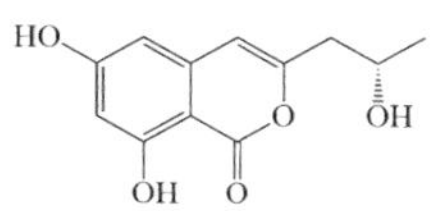

6,8 dihydroxy (3) (2 oxopropyl) isocoumarin **274**

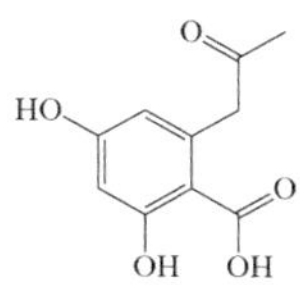

6,8 dihydroxy 3 [(2S) 2 hydroxypropyl] isocoumarin **275**

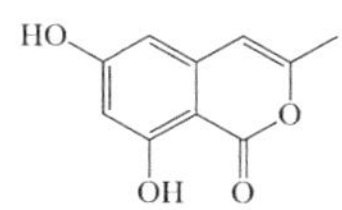

2,4 dihydroxy 6 (2 oxopropyl) benzoic acid **276**

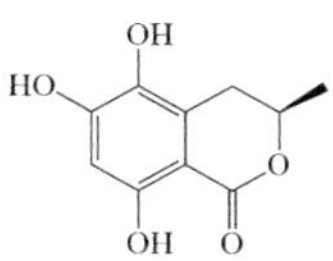

6,8 dihydroxy 3 methylisocoumarin **277**

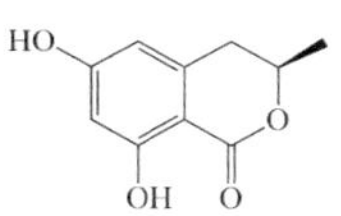

5,6,8 trihydroxy 3(R) methyl 3,4 dihydroisocoumarin **278**

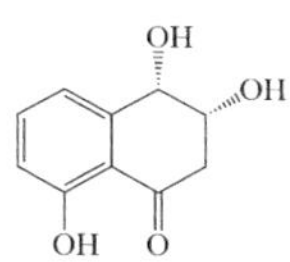

6,8 dihydroxy 3(R) methyl 3,4 dihydroisocoumarin **279**

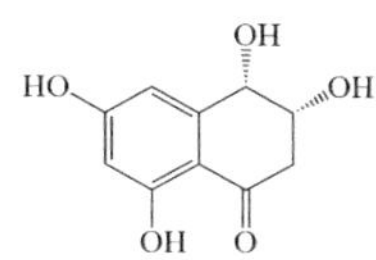

(3R,4S) 3,4,8 trihydroxy 3,4 dihydro 1(2H) naphthalenone **280**

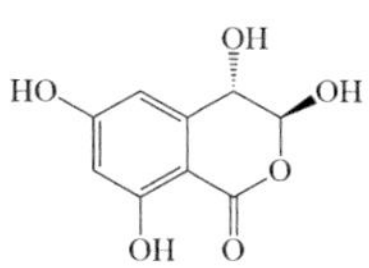

(3S,4S) 3,4,6,8 tetrahydroxy 3,4 dihydro 1(2H) naphthalenone **281**

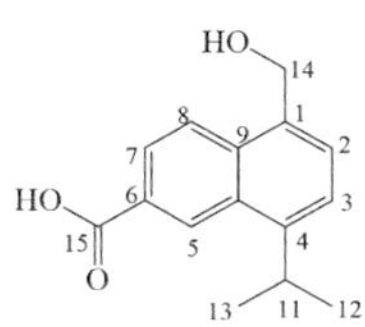

Pericoterpenoid A **282**

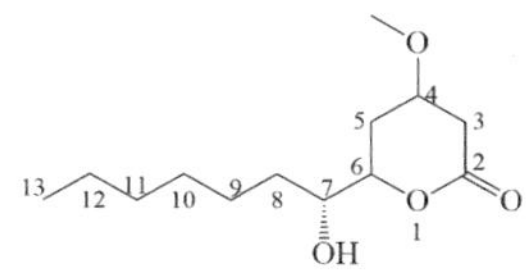

Nodulisporipyrones A **283**

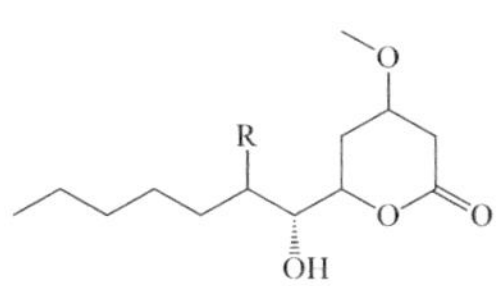

284 Nodulisporipyrones B R ◀OH
285 Nodulisporipyrones C R ⋯OH

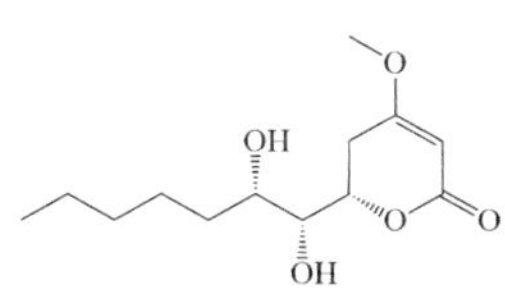

Nodulisporipyrones D **286**

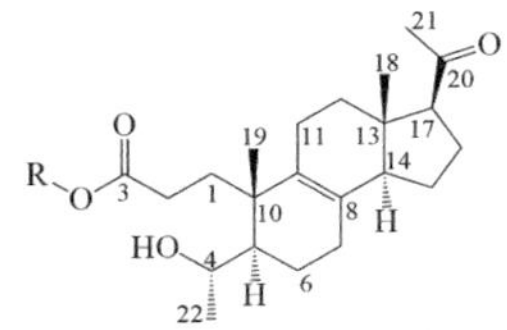

287 R = H Nodulisporisteroids C
288 R = Me Nodulisporisteroids D

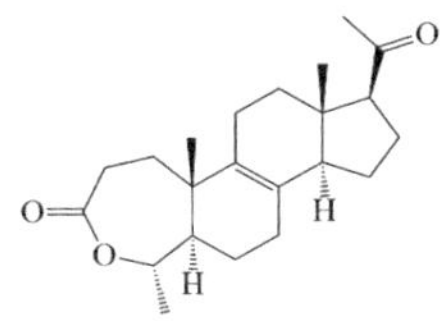

Nodulisporisteroids E **289**

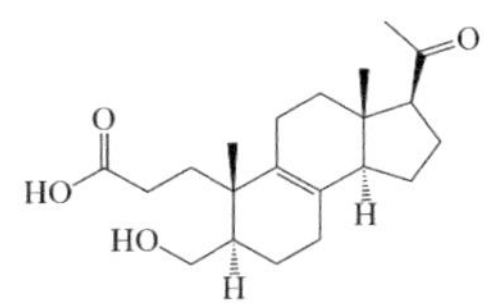

Nodulisporisteroids F **290**

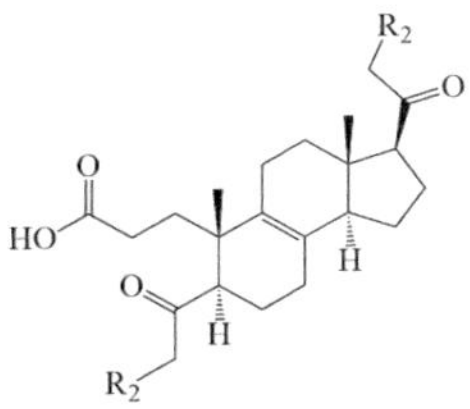

291 R_1 = H, R_2 = H Nodulisporisteroids G
292 R_1 = H, R_2 = OH Nodulisporisteroids H
293 R_1 = OH, R_2 = H Nodulisporisteroids I

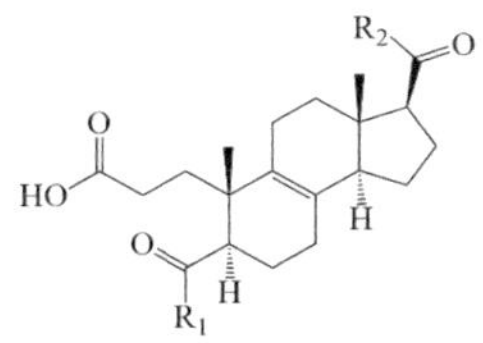

294 R_1 = OH, R_2 = Me Nodulisporisteroids J
295 R_1 = Me, R_2 = OH Nodulisporisteroids K

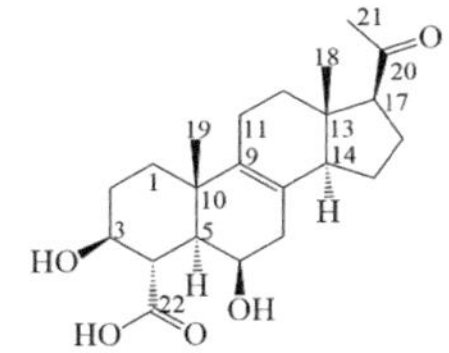

Nodulisporisteroids L **296**

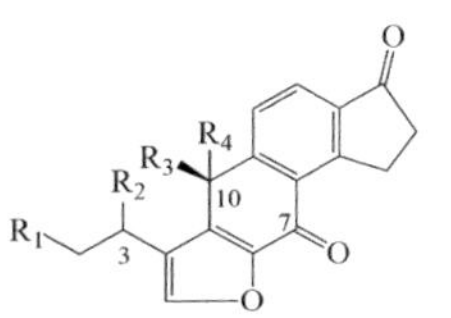

	R_1	R_2	R_3	R_4
297	CH_2OH	····OH	OH	CH_3
298	CH_2OH	H	OH	CH_3
299	OH	H	OH	CH_3
300	CH_2OH	····OH	CH_3	OH

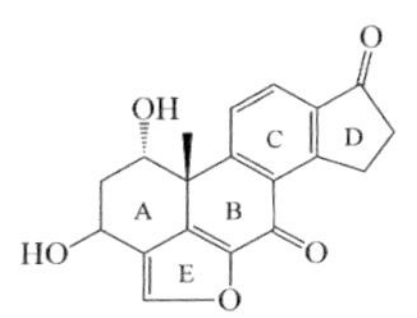

Nodulisporiviridins E **301**

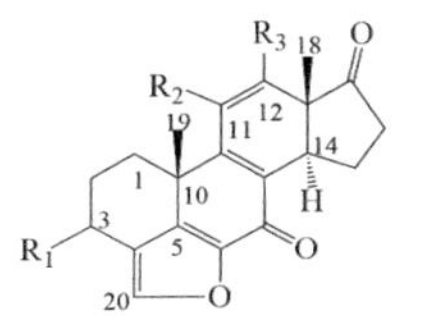

	R_1	R_2	R_3
302	◀OH	◀OH	H
303	O::::	◀OH	H
304	◀OH	H	◀OH

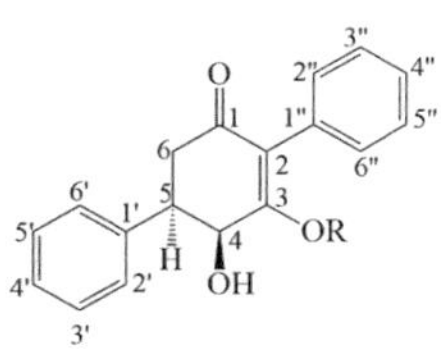

305 Cucurbitarin A R = Me
306 Cucurbitarin B R = H

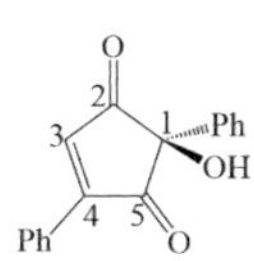

Cucurbitarin D **308**

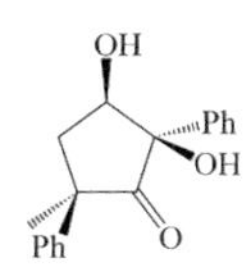

Cucurbitarin E **309**

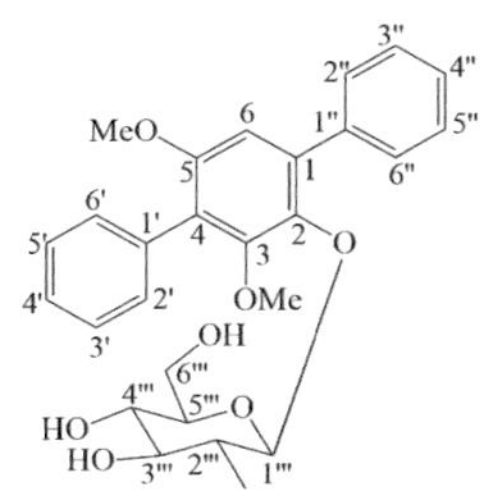

Cucurbitarin C **307**

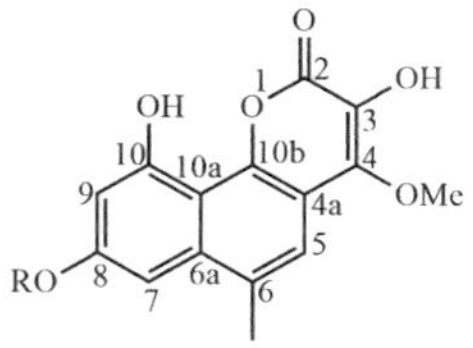

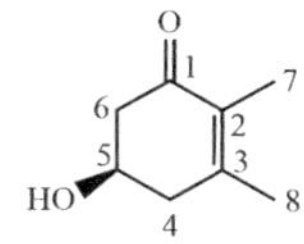

310 3,10-dihydroxy-4,8-dimethoxy-6-methylbenzocoumarin R = Me

311 3,8,10-trihydroxy-4-methoxy-6-methylbenzocoumarin R = H

(5*R*)-5-hydroxy-2,3-dimethylcyclohex-2-en-1-one **312**

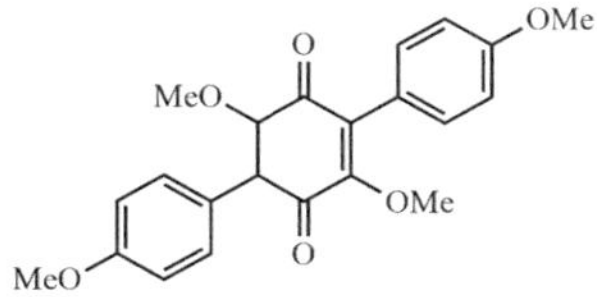

2,5-dimethoxy-3,6-bis(4-methoxypheny1)-1,4-benzoquinone **313**

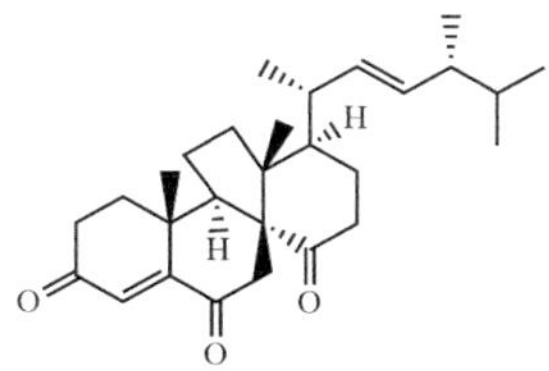

Dankasterone A **314**

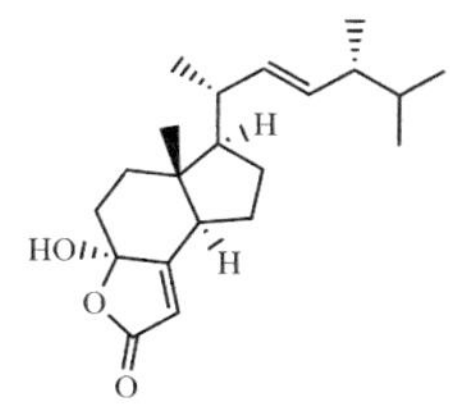

(17R)-4-hydroxy-17-methylincisterol **315**

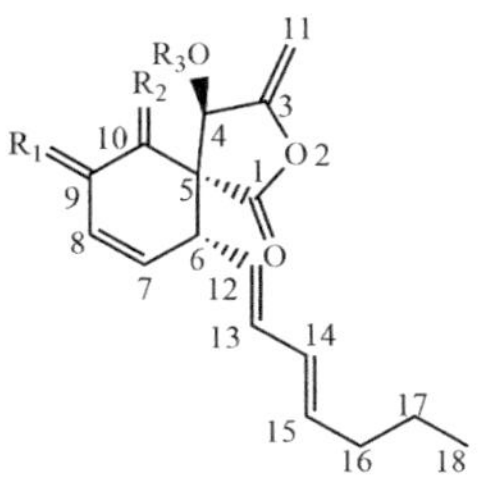

	R_1	R_2	R_3
316 Oxaspirol A	α–H, β–OH	α–OH, β–H	H
318 Oxaspirol C	α–H, β–OH	α–H, β–OH	H

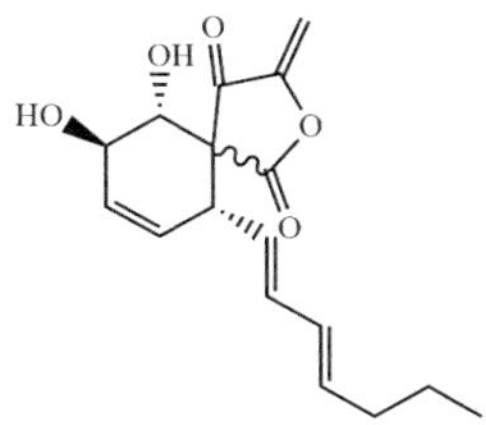

Oxaspirol B **317**

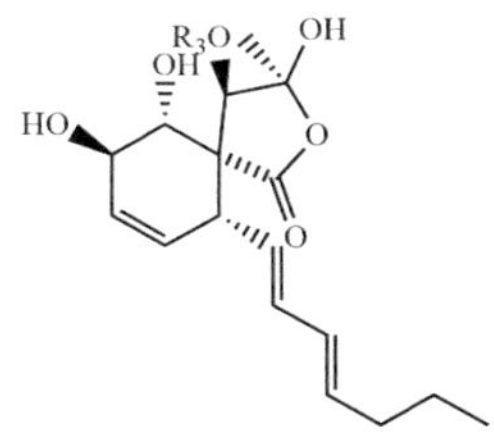

Oxaspirol D **319**

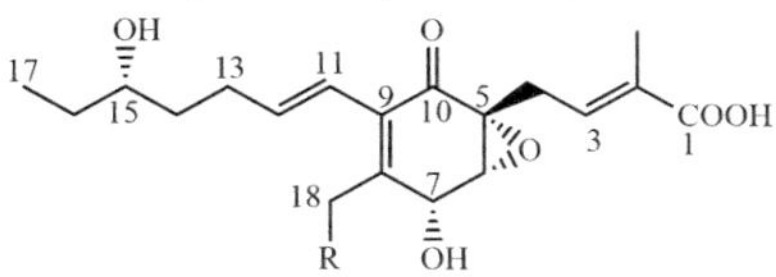

320 Ambuic acid R = OH

321 Ambuic acid R = OAc

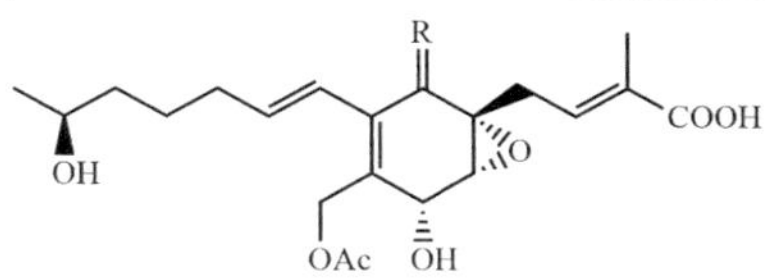

322 Ambuic acid R = αOH

323 Ambuic acid R = ═O

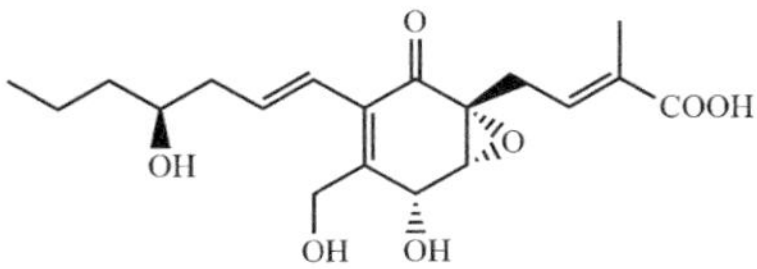

Ambuic acid **324**

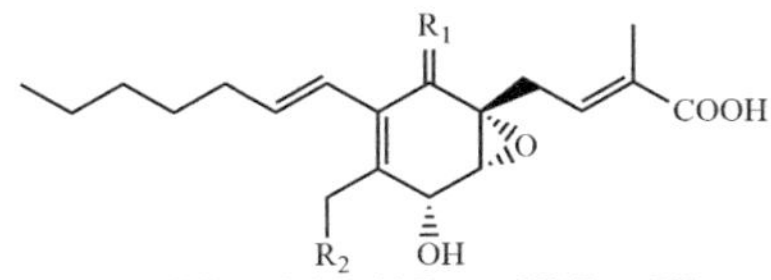

325 Ambuic acid $R_1 = \alpha OH$, $R_2 = OH$
326 Ambuic acid $R_1 = \alpha OH$, $R_2 = OAc$
327 Ambuic acid $R_1 = {=}O$, $R_2 = OH$

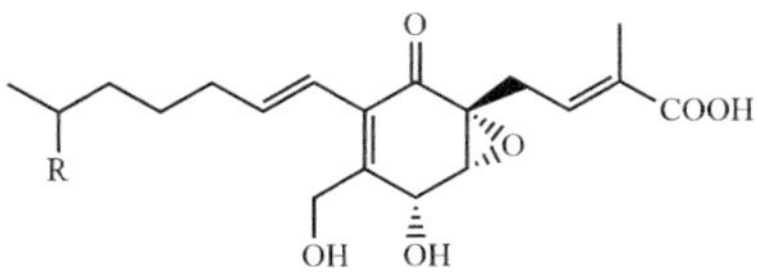

328 Ambuic acid R = βOH
329 Ambuic acid R = =O

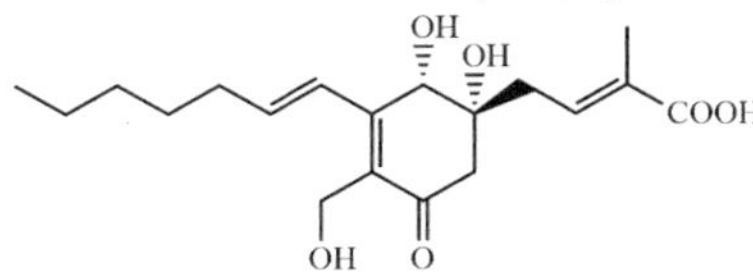

330 Ambuic acid

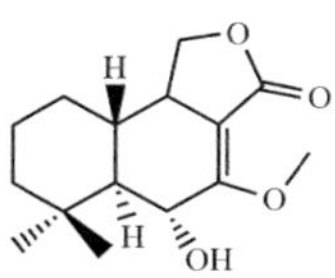

Chaetothyrins A **331**

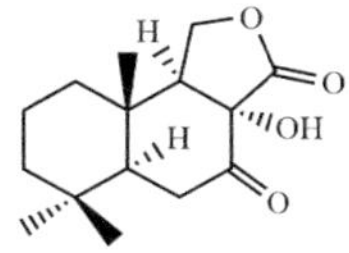

Chaetothyrins B **332**

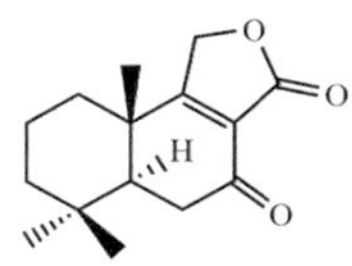

Chaetothyrins C **333**

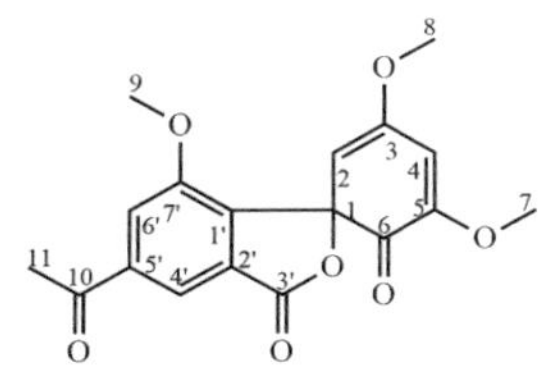

5'-acetyl-3,5,7'-trimethoxy-3'H-spiro[cyclohexa [2,4]diene-1,1'-isobenzofuran]-3',6-dione **334**

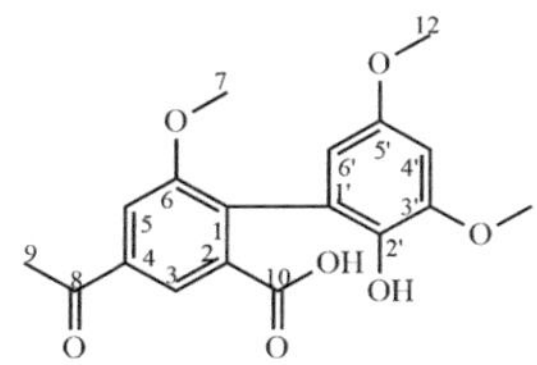

4-acetyl-2'-hydroxy-3',5',6-trimethoxy biphenyl-2-carboxylic acid **335**

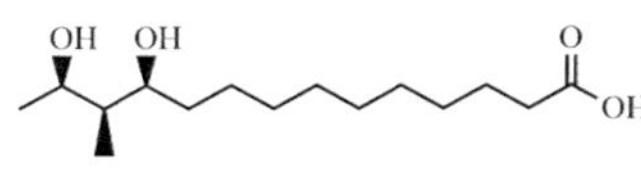

(11*S*,12*S*,13*R*)-11,13-dihydroxy-12-methyltetradecanoic acid **336**

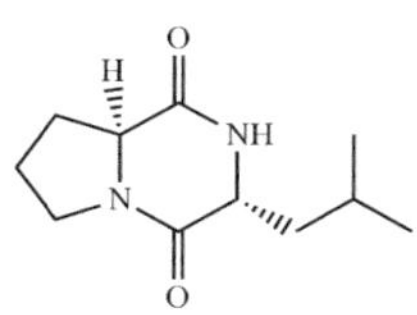

Cyclo-(S-Pro-R-Leu) **337**

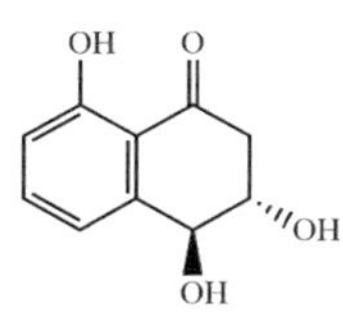

3,4-dihydro-3,4,8-trihydroxy (2H)naphthalenone **338**

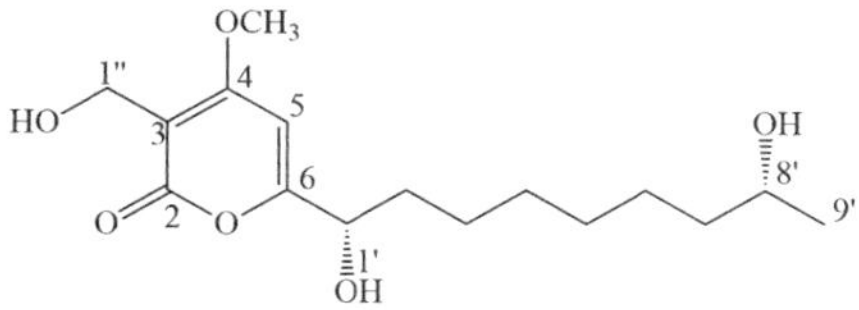

Dothideopyrones E **339**

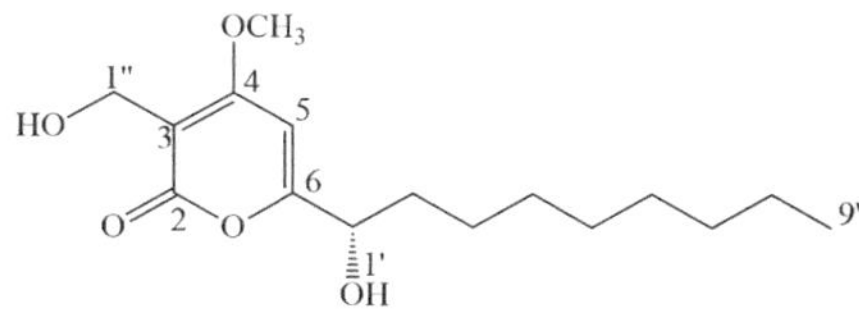

Dothideopyrones F **340**

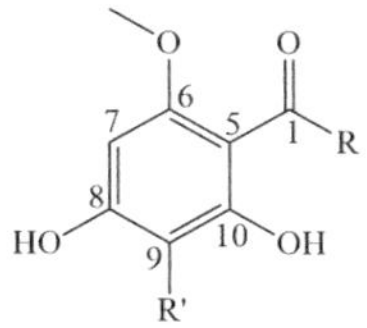

Phomalichenones A **341** R = CHCH(*E*)CH_3, R' = CH_2CH_3
Phomalichenones B **342** R = $(CH_2)_3OH$, R' = CH_2CH_2OH
(2,4-dihydroxy-3-(2-hydroxyethyl)-6-methoxyphenyl)-3 hydroxybutan-1-one **345** R = $CH_2CHOHCH_3$, R' = CH_2CH_2OH
(E)-1-(2,4- dihydroxy-3-(2-hydroxyethyl)-6-methoxyphenyl)but-2-en-1-one **346** R = CHCH(*E*)CH_3, R' = CH_2CH_2OH
Phomalone **347** R = $CH_2CH_2CH_3$, R' = CH_2CH_2OH
Deoxyphomalone **348** R = $CH_2CH_2CH_3$, R' = CH_2CH_3

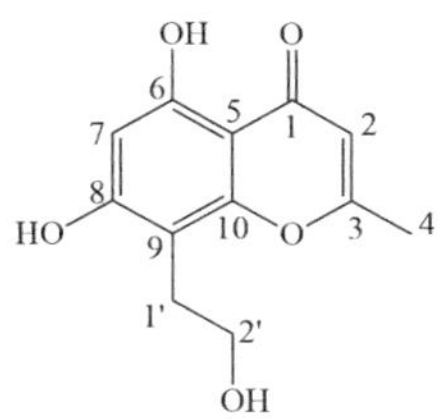

Phomalichenones C **343**

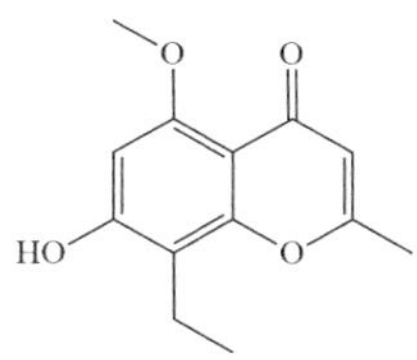

Phomalichenones D **344**

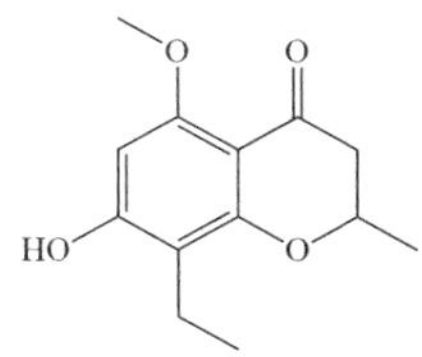

8-ethyl-7-hydroxy-5-methoxy-2-methylchroman-4-one **349**

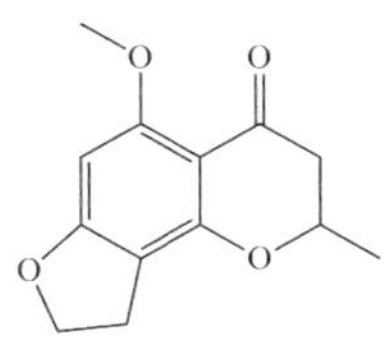

LL-D253Y **350**

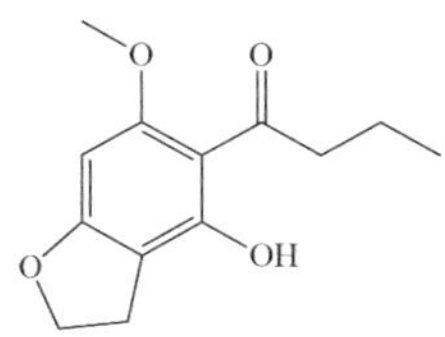

4-hydroxy-6-methoxy-5-(1'-oxobutyl) benzodihydrofuran **351**

References

Boonpragob K, Crittenden PD, Lumbsch TH (2012) Lichens: from genome to ecosystems in a changing world. MycoKeys 6:1–2

Chang W, Zhang M, Li Y et al (2015) Lichen endophyte derived pyridoxatin inactivates *Candida* growth by interfering with ergosterol biosynthesis. Biochim Biophys Acta 1850(9):1762–1771

Chen GD, Li YJ, Gao H et al (2012) New azaphilones and chlorinated phenolic glycosides from *Chaetomium elatum* with caspase-3 inhibitory activity. Planta Med 78:1683–1689

Chen GD, Chen Y, Gao H et al (2013) Xanthoquinodins from the endolichenic fungal strain *Chaetomium elatum*. J Nat Prod 76:702–709

Chen GD, Bao YR, Huang YF et al (2014) Three pairs of variecolortide enantiomers from *Eurotium* sp. with caspase-3 inhibitory activity. Fitoterapia 92:252–259
Demain AL (1999) Pharmaceutically active secondary metabolites of microorganisms. Appl Microbiol Biotechnol 52:455–463
Ding G, Li Y, Fu S et al (2009) Ambuic acid and torreyanic acid derivatives from the endolichenic fungus *Pestalotiopsis* sp. J Nat Prod 72(1):182–186
Dou Y, Wang X, Jiang D et al (2014) Metabolites from *Aspergillus versicolor*, an endolichenic fungus from the lichen *Lobaria retigera*. Drug Discov Ther 8(2):84–88
Dreyfuss MM, Chapela IH (1994) Potential of fungi in the discovery of novel, low-molecular weight pharmaceuticals. In: Gullo VP (ed) The discovery of natural products with therapeutic potential. Butterworth-Heinmann, Boston, pp 49–80
Feuerer T, Hawksworth DL (2007) Biodiversity of lichens, including a world-wide analysis of checklist data based on Takhtajan's floristic regions. Biodivers Conserv 16:85–98
Firakova S, Sturdikova M, Muckova M (2007) Bioactive secondary metabolites produced by microorganisms associated with plants. Biologia 62(3):251–257
Gloer JB (1997) Applications of fungal ecology in the search for new bioactive natural products. In: Wicklow DT, Soderstrom BE (eds) The mycota, Environmental and microbial relationships, vol 4. Springer, New York, pp 249–268
Hanson JR (2008) Chemistry of fungi. RSC Publishing, Cambridge, pp 231
Hassan RS, Strobel GA, Booth E et al (2012) Modulation of volatile organic compound formation in the mycodiesel producing endophyte- *Hypoxylon* sp. C1-4. Microbiology 158:464–473
Hawksworth DL (1991) The fungal dimension of biodiversity: magnitude, significance, and conservation. Mycol Res 95:641–655
Hawksworth DL (2001) The magnitude of fungal diversity: the 1.5 million species estimate revisited. Mycol Res 105:1422–1432
He JW, Chen GD, Gao H et al (2012) Heptaketides with antiviral activity from three endolichenic fungal strains *Nigrospora* sp., *Alternaria* sp. and *Phialophora* sp. Fitoterapia 83(6):1087–1091
Heywood VH (ed) (1995) Global biodiversity assessment. Cambridge University Press, Cambridge
Huang YF, Li XX, Chen GD et al (2012) A new diphenyl ether from an endolichenic fungal strain, I sp. Mycosystema 31:769–774
Jiao Y, Li G, Wang HY et al (2015) New metabolites from endolichenic fungus Pleosporales sp. Chem Biodivers 12(7):1095–1104
Keller NP, Turner G, Bennett JW (2005) Fungal secondary metabolism-from biochemistry to genomics. Nat Rev Microbiol 3:937–947
Kim KH, Beemelmanns C, Murillo C et al (2014) Naphthalenones and isocoumarins from a Costa Rican fungus Xylariaceae sp. CR1546C. J Chem Res 38(12):722–725
Kim GS, Ko W, Kim JW et al (2018a) Bioactive α-Pyrone derivatives from the endolichenic fungus dothideomycetes sp. EL003334. J Nat Prod 81:1084–1088
Kim JW, Ko W, Kim E et al (2018b) Anti-inflammatory phomalichenones from an endolichenic fungus *Phoma* sp. J Antibiot 71:753–756
Li YJ, Li XX, Chen GD et al (2012a) J Shenyang Pharm Univ 29:678–701
Li G, Wang H, Zhu R et al (2012b) Phaeosphaerins A-F, cytotoxic perylenequinones from an endolichenic fungus *Phaeosphaeria* sp. J Nat Prod 75(2):142–147
Li W, Li XB, Li L et al (2015a) α-Pyrone derivatives from the endolichenic fungus *Nectria* sp. Phytochem Lett 12:22–26
Li XB, Li L, Zhu RX et al (2015b) Tetramic acids and pyridone alkaloids from the endolichenic fungus *Tolypocladium cylindrosporum*. J Nat Prod 78(9):2155–2160
Li XB, Zhou YH, Zhu RX et al (2015c) Identification and biological evaluation of secondary metabolites from the endolichenic fungus *Aspergillus versicolor*. Chem Biodivers 12:575–592
MacMillan J (2002) Occurrence of gibberellins in vascular plants, fungi and bacteria. J Plant Growth Regul 20:387–442
McAlpine JB, Bachmann BO, Piraee M et al (2005) Microbial genomics as a guide to drug discovery, structural elucidation: ECO02301, a novel antifungal agent, as an example. J Nat Prod 68:493–496

Paranagama PA, Wijeratne EMK, Burns AM et al (2007) Heptaketides from *Corynespora* sp. inhabiting the cavern beard lichen, *Usnea cavernosa*: first report of metabolites of an endolichenic fungus. J Nat Prod 70(11):1700–1705

Porras-Alfaro A, Bayman P (2011) Hidden fungi, emergent properties: endophytes and microbiomes. Annu Rev Phytopathol 49:291–315

Proudfoot JR (2002) Drugs, leads and drug-likeness: an analysis of some recently launched drugs. Bioorg Med Chem Lett 12:1647–1650

Samanthi KAU, Wickramaarachchi S, Wijeratne EMK et al (2015) Two new antioxidant active polyketides from *Penicillium citrinum*, an endolichenic fungus isolated from *Parmotrema* species in Sri Lanka. J Natl Sci Found Sri Lanka 43(2):119–126

Schulz B, Boyle C, Draeger S et al (2002) Endophytic fungi: a source of novel biologically active secondary metabolites. Mycol Res 106:996–1004

Staley JT, Castenholz RW, Colwell RR et al (1997) The microbial world: foundation of the biosphere. American Academy of Microbiology, Washington DC, pp 32

Stierle A, Strobel G, Stierle D (1993) Taxol and taxane production by *Taxomyces andreanae*, an endophytic fungus of Pacific yew. Science 260(9):214–216

Strobel GA (2002) Microbial gifts from the rainforest. Can J Phytopathol 24:14–20

Strobel G (2006) Harnessing endophytes for industrial microbiology. Curr Opin Microbiol 9:240–244

Strobel G, Daisy B (2003) Bioprospecting for microbial endophytes and their products. Microbiol Mol Biol Rev 67:491–502

Strobel GA, Hess WM, Ford E et al (1996) Taxol from fungal endophytes and the issue of biodiversity. J Ind Microbiol 17:417–423

Tan RX, Zou WX (2001) Endophytes: a rich source of functional metabolites. Nat Prod Rep 18:448–459

Taylor TN, Taylor EL (2000) The rhynie chert ecosystem: a model for understanding fungal interactions. In: Bacon CW, White JF (eds) Microbial endophytes. Marcel Decker, New York, pp 31–48

Wang Y, Niu S, Liu S et al (2010a) The first naturally occurring thiepinols and thienol from an endolichenic fungus *Coniochaeta* sp. Org Lett 12(21):5081–5083

Wang Y, Zheng Z, Liu S et al (2010b) Oxepinochromenones, furochromenone, and their putative precursors from the endolichenic fungus *Coniochaeta* sp. J Nat Prod 73(5):920–924

Wang QX, Bao L, Yang XL et al (2012) Polyketides with antimicrobial activity from the solid culture of an endolichenic fungus *Ulocladium* sp. Fitoterapia 83(1):209–214

Wang QX, Bao L, Yang XL et al (2013a) Tricycloalternarenes F–H: three new mixed terpenoids produced by an endolichenic fungus *Ulocladium* sp. using OSMAC method. Fitoterapia 85(1):8–13

Wang QX, Bao L, Yang XL et al (2013b) Ophiobolins P-T, five new cytotoxic and antibacterial sesterterpenes from the endolichenic fungus *Ulocladium* sp. Fitoterapia 90:220–227

Wijeratne EMK, Bashyal BP, Gunatilaka MK et al (2010) Maximizing chemical diversity of fungal metabolites: biogenetically related heptaketides of the endolichenic fungus *Corynespora* sp. J Nat Prod 73(6):1156–1159

Wijeratne EMK, Bashyal BP, Liu MX et al (2012) Geopyxins A-E, ent –Kaurane diterpenoids from endolichenic fungal strains *Geopyxis* aff. *majalis* and *Geopyxis* sp. AZ0066: structure-activity relationships of geopyxins and their analogues. J Nat Prod 75(3):361–369

Wijeratne EMK, Gunaherath GMKB, Chapla VM et al (2016) Oxaspirol B with p97 inhibitory activity and other Oxaspirols from *Lecythophora* sp. FL1375 and FL1031, endolichenic fungi inhabiting *Parmotrema tinctorum* and *Cladonia evansii*. J Nat Prod 79(2):340–352

Wu W, Dai H, Bao L et al (2011) Isolation and structural elucidation of proline-containing cyclopentapeptides from an endolichenic *Xylaria* sp. J Nat Prod 74(5):1303–1308

Wu YH, Chen GD, Wang CX et al (2015) Pericoterpenoid A, a new bioactive cadinane-type sesquiterpene from *Periconia* sp. J Asian Nat Prod Res 17(6):671–675

Xiong H, Xiao GK, Chen GD et al (2014) Sporormiellin A, the first tetrahydrofuran-fused furochromone with an unprecedented tetracyclic skeleton from *Sporormiella minima*. RSC Adv 4(46):24295–24299
Yang F, Chen GD, Gao H et al (2012) Two new naphthalene derivatives from an endolichenic fungal strain *Scopulariopsis* sp. J Asian Nat Prod Res 14(11):1059–1063
Yang BJ, Chen GD, Li YJ et al (2016) A new xanthone glycoside from the endolichenic fungus *Sporormiella irregularis*. Molecules 21(6). https://doi.org/10.3390/molecules21060764
Yang Y, Bae WK, Nam SJ et al (2018) Acetonic extracts of the endolichenic fungus EL002332 isolated from *Endocarpon pusillum* exhibits anticancer activity in human gastric cancer cells. Phytomedicine 1(40):106–115
Ye F, Chen GD, He JW et al (2013) Xinshengin, the first altenusin with tetracyclic skeleton core from *Phialophora* spp. Tetrahedron Lett 54(34):4551–4554
Yuan C, Wang HY, Wu CS et al (2013) Austdiol, fulvic acid and citromycetin derivatives from an endolichenic fungus, *Myxotrichum* sp. Phytochem Lett 6(4):662–666
Yuan C, Li G, Wu CS et al (2015) A new fatty acid from the endolichenic fungus *Massarina* sp. Chem Nat Compd 51(3):415–417
Yuan C, Guo YH, Wang HY et al (2016) Allelopathic polyketides from an endolichenic fungus *Myxotrichum* sp. by using OSMAC strategy. Sci Rep 6:19350. https://doi.org/10.1038/srep19350
Yuan C, Ding G, Wang HY et al (2017) Polyketide-terpene hybrid metabolites from an endolichenic fungus *Pestalotiopsis* sp. Biomed Res Int 2017:1–10. https://doi.org/10.1155/2017/6961928
Zhang F, Liu S, Lu X et al (2009) Allenyl and alkynyl phenyl ethers from the endolichenic fungus *Neurospora terricola*. J Nat Prod 72(10):1782–1785
Zhang F, Li L, Niu S et al (2012) A thiopyranchromenone and other chromone derivatives from an endolichenic fungus, *Preussia africana*. J Nat Prod 75:230–237
Zhang K, Ren J, Ge M et al (2014) Mono- and bis-furanone derivatives from the endolichenic fungus *Peziza* sp. Fitoterapia 92:79–84
Zhao H, Wang GQ, Tong XP et al (2014) Diphenyl ethers from *Aspergillus* sp. and their anti-Aβ42 aggregation activities. Fitoterapia 98:77–83
Zhao Q, Chen GD, Feng XL et al (2015a) Nodulisporiviridins A-H, bioactive viridins from *Nodulisporium* sp. J Nat Prod 78(6):1221–1230
Zhao Q, Wang CX, Yu Y et al (2015b) Nodulisporipyrones A-D, new bioactive a-pyrone derivatives from *Nodulisporium* sp. J Asian Nat Prod Res 17(5):567–575
Zhao Q, Wang GQ, Chen GD et al (2015c) Nodulisporisteroids C-L, new 4-methyl-progesteroid derivatives from *Nodulisporium* sp. Steroids 102:101–109
Zheng QC, Chen GD, Kong MZ et al (2013) Nodulisporisteriods A and B, the first 3,4-seco-4-methyl-progesteroids from *Nodulisporium* sp. Steroids 78(9):896–901
Zheng QC, Kong MZ, Zhao Q et al (2014) Chaetoglobosin Y, a new cytochalasan from *Chaetomium globosum*. Fitoterapia 93:126–131
Zhou YH, Zhang M, Zhu RX et al (2016a) Heptaketides from an endolichenic fungus *Biatriospora* sp. and their antifungal activity. J Nat Prod 79(9):2149–2157
Zhou YH, Li XB, Zhang JZ et al (2016b) Three new drimane-type sesquiterpenoids, chaetothyrins A–C, from an endolichenic fungus Chaetothyriales sp. J Asian Nat Prod Res 18(5):409–414

Endolichenic Fungi: A Case Study from Uttarakhand

Abstract Uttarakhand have prolific lichen diversity and a lot of studies regarding lichen diversity were preformed but studies on endolichenic fungi started in the recent past not only in Uttarakhand but in India also. The authors started working on endolichenic fungi in the Kumaun region of Uttarakhand. We studied the seasonal variation, the similarities between the diversity of endolichenic fungi and the endophytes of host tree on which the lichen thallus was growing and diversity of secondary metabolites isolated from endolichenic fungi and their antimicrobial potential. This chapter documents a case study on endolichenic fungi from Uttarakhand and reveals the presence of 42 species belonging to Dothediomycetes, Eurotiomycetes, Leotiomycetes, Sordariomycetes and Zygomycetes. Ten species of fungi (*Acremonium lichenicola*, *Bipolaris australiensis*, *Nigrospora sphaerica*, *Papulospora* sp., *Pestalotiopsis maculans*, *Rhizoctonia* sp., *Sordaria fimicola*, *Spegazzinia tessarthra*, *Trichophyton roseum* and *Xylaria hypoxylon*) were documented originally as endolichenic fungi in this study. The study also revealed that some endolichenic fungi are specific (*Spegazzinia tessarthra*, *Nigrospora oryzae*, *Pestalotiopsis maculans*, *Sordaria fimicola*, *Rhizoctonia* sp.) while others are generalized (*Alternaria alternata*, *Aspergillus flavus* and *Fusarium solani*). Among crop pathogens *X. campestris* was found to be the most susceptible and *R. solanacearum* the most resistant pathogen. Among human pathogens *E. coli* and *S. typhimurium* were found to be equally susceptible while *P. aeruginosa* the most resistant pathogen. Among all pathogens, extract of *Aspergillus niger* came out as a broad spectrum antibiotic. Three strains of mycelia sterilia (A01, D31 and E48) showed moderate antimicrobial activity.

This study reveals the hidden diversity of endolichenic fungi and gives a natural alternative to synthetic drugs available in market. The present investigation is an attempt in this path, and our investigation proposes that endolichenic fungi could be a prospective supply of antimicrobial agents.

Despite rich biodiversity of lichens in India [2714 species (Sinha et al. 2018)] and their diverse growth forms, ranging from leprose to crustose to squamulose to foliose to fruticose, the endolichenic fungi residing inside the lichen thallus are likely to be the least well known biological group of fungi in India. This is because of the

M. Tripathi, Y. Joshi, *Endolichenic Fungi: Present and Future Trends*,
https://doi.org/10.1007/978-981-13-7268-1_6

comparably poor attention that these fungi have received from mycologists, including lichenologists. The scientific knowledge of endolichenic fungi in India started recently when Suryanarayanan et al. (2005) for the first time isolated endophytic fungi from some lichens. This extremely important work has to a large extent been the baseline for our current work (particularly their diversity and distribution in Kumaun Himalaya) and most of our findings have already been published (Tripathi et al. 2014a, b, c; Triapthi and Joshi 2015).

A lot of information pertaining to various aspects of lichens of Uttarakhand *viz.* floristic diversity, revisionary studies, bio-prospection studies and pollution monitoring studies, are available, but studies pertaining to endolichenic fungi are very few and hence an attempt is made by the authors for increasing the interest and awareness amongst academicians in endolichenic fungi in India, by studying the endolichenic fungi associated with some macrolichens of Kumaun Himalaya.

Though the authors isolated endolichenic fungi from 22 macrolichens of Almora and Champawat districts, but they focused their study on the endolichenic fungi associated with four macrolichen species (*Bulbothrix setschwanensis*, *Flavoparmelia caperata*, *Flavopunctelia flaventior* and *Parmotrema reticulatum*) collected from Syahi Devi region of Almora district, Kumaun Himalaya because of following reasons:

(1) isolating endolichenic fungi chiefly depends on the methodology applied and diversity of endophytes is directly proportional to the number of lichen segments incubated i.e. more the lichen samples we incubate, more the fungi we get. Hence, time and fund constraints made it impossible to screen each and every lichen species from entire Kumaun Himalaya, which is stretched in an area of 53,483 km^2.
(2) to avail better results the collected lichen samples should be processed within 24 h of its collection, and this could be possible if the site is near to proximity of working place, and.
(3) to know seasonal variations and host specificity, the site should be easily accessible, hence authors selected Syahi Devi forest which is just 36 km from the headquarter.

The macrolichens were collected with the help of chisel and hammer along with their substratum from different forest sites of Almora and Champawat districts. The samples were placed in sterilized paper bags and processed within 24 h of their collection. The macrolichens were identified by following Awasthi (2007) and for knowing the endolichenic diversity macrolichens were surface sterilized by using the modified protocol of Suryanarayanan et al. (2005). Macrolichen samples were rinsed carefully in running tap water to wash away the unwanted materials. Then the samples were dipped in petri plate containing sterile double distilled water and bryophytes/mosses were removed and samples were then again washed in sterile double distilled water 20 times until all the other visible contaminants get removed. Then the samples were subjected to chemical surface sterilization by dipping them in 30% Hydrogen peroxide (H_2O_2) for 30 s, followed by 4% Sodium hypochlorite (NaOH) for 30 s and finally immersing them in 75% ethanol for 30 s. After chemical surface sterilization the samples were washed in double distilled autoclaved water twice and dried under aseptic conditions and were cut into 0.5 × 0.5 cm pieces.

The surface sterilized samples were placed on petri dishes containing Potato Dextrose Agar (PDA) supplemented with 150 mg/l streptomycin and sealed using Parafilm™. Efficiency of surface sterilization was tested by plating out 500 μl of the last rinsing water from the sterilization procedure and tissue imprint on fresh PDA plates (Schulz et al. 1993). Plates were incubated at 25 ± 1 °C until fungal growth was initiated. The growing tips of fungal mycelia were transferred to new PDA plates for obtaining pure culture and the pure cultures were then examined periodically. After 15 days, those endophytes that had grown on the media were identified using published floras and the ones which did not sporulate (sterile cultures) were termed as mycelia sterilia. The identification of endolichenic fungi was carried out by their colonial morphology, microscopic observations and nature of the spores. Temporary mounts of the fungi were made in lactophenol cotton blue by using Scotch Tape technique (Harris 2000) and were identified using relevant keys and taxonomic notes (Gilman 1967; Ellis 1971, 1976; Sutton 1980; Chowdhry 2000). All the cultures are deposited in the herbarium of Kumaun University, Almora.

The study revealed the presence of 42 species of endolichenic fungi (excluding Mycelia sterilia) belonging to 31 genera (Tables 1 and 2) encompassing members of Dothediomycetes, Eurotiomycetes, Leotiomycetes, Sordariomycetes and Zygomycetes. Out of these 42 endolichenic fungi, 10 fungal species (*Acremonium lichenicola*, *Bipolaris australiensis*, *Nigrospora oryzae*, *Papulospora* sp., *Pestalotiopsis maculans*, *Rhizoctonia* sp., *Sordaria fimicola*, *Spegazzinia tessarthra*, *Trichophyton roseum* and *Xylaria hypoxylon*) are reported for the first time across the world as endolichenic fungi. Besides this, the study also revealed that some species of fungi are lichen specific, i.e. restricted to a particular lichen (*Spegazzinia tessarthra*, *Nigrospora oryzae*, *Pestalotiopsis maculans*, *Sordaria fimicola*, *Rhizoctonia* sp.), while some species are more generalized and are reported from several or all target lichen species (*Alternaria alternata, Aspergillus flavus* and *Neocosmospora solani*).

The present study revealed the occurrence of 268 isolates of endolichenic fungi belonging to 14 morphospecies in winter season (excluding mycelia sterilia) (Table 3) and 318 isolates belonging to 09 morphospecies in summer season (excluding mycelia sterilia) (Table 4) isolated from 04 macrolichen species (Figs. 1, 2, 3, 4 and 5). The maximum numbers of endolichenic fungal species (15) were isolated from the thallus of *P. reticulatum* in both the seasons (Tables 3 and 4). Mycelia sterilia was a major proportion of endolichenic fungi isolated in this study for all the studied lichen species in both seasons (Figs. 4 and 5).

In winter season among all four tested macrolichen species *Parmotrema reticulatum* bears the most diverse endolichenic fungal population (11 species), while *Flavoparmelia caperata* was the least diverse (03 species). In summer season, *Bulbothrix setschwanensis* proved itself to be the best host by harboring the highest number of endolichenic fungi (08), while rest of the three macrolichen species viz. *Flavoparmelia caperata*, *Parmotrema reticulatum* and *Flavopunctelia flaventior* (06, 06 and 05, respectively) were almost harboring similar number of species.

Table 1 Endolichenic fungi isolated from some macrolichens of Almora district

Lichen species	Family	Endolichenic fungi
Bulbothrix meizospora	Parmeliaceae	*Alternaria alternata*, *Aspergillus flavus*, *Cylindrosporium* sp., *Gilmaniella humicola*, mycelia sterilia, *Neocosmospora solani* (= *Fusarium solani*), *Penicillium* sp.
Flavoparmelia caperata	Parmeliaceae	*A. alternata*, *A. flavus*, *N. solani*, mycelia sterilia
Heterodermia flabellata	Physciaceae	*A. alternata*, *A. flavus*, *A. niger*, *Curvularia australiensis* (= *Bipolaris australiensis*), *N. solani*, *Spegazzinia tessarthra*, *Trichoderma harzianum*
Heterodermia hypochraea	Physciaceae	*A. alternata*, *A. flavus*, *A. niger*, *N. solani*, *Papulospora* sp.
Leptogium burnetiae	Collemataceae	*A. alternata*, *A. flavus*, *N. solani*, *G. humicola*
Parmotrema thomsonii	Parmeliaceae	*Acremonium* sp., *A. alternata*, *A. flavus*, *A. niger*, *N. solani*, *Nigrospora oryzae* (= *N. sphaerica*), *Pestalotiopsis* sp., *T. harzianum*
Parmotrema crinitum	Parmeliaceae	*A. alternata*, *A flavus*, *N. solani*, mycelia sterilia, *T. harzianum*
Parmotrema grayanum	Parmeliaceae	*A. alternata*, *A. flavus*, *G. humicola*, *N. solani*, *Trichophyton roseum*
Parmotrema nilgherrense	Parmeliaceae	*A. alternata*, *A. flavus*, *Chaetomella* sp., *Cladosporium* sp., *G. humicola*, *N. solani*, mycelia sterilia
Parmotrema praesorediosum	Parmeliaceae	*A. alternata*, *A. flavus*, *Cladosporium* sp., *N. solani*
Parmotrema reticulatum	Parmeliaceae	*Acremonium lichenicola*, *A. alternata*, *A. flavus*, *N. solani*, *Nigrospora oryzae*, *Mucor racemosus*, *Papulospora* sp., *Penicillium* sp., *Pestalotiopsis maculans*, *Sordaria fimicola*, *Xylaria hypoxylon*
Physcia dilatata	Physciaceae	*A. alternata*, *A. flavus*, *A. niger*, *C. australiensis*, *Cladosporium* sp., *N. solani*, *T. harzianum*
Punctelia subrudecta	Parmeliaceae	*G. humicola*, *N. solani*, *Rhizoctonia* sp.
Usnea sp.	Parmeliaceae	*N. solani*, *T. harzianum*

In winters, members of Eurotiomycetes were more frequently observed, while in summers, members of Sordariomycetes were found in abundance. Members of Leotiomycetes were found only in summers, whereas members of Dothediomycetes were common in both the seasons. In contrast to winter where members of Zygomycetes were present only in one lichen (*P. reticulatum*), in summer they were reported from all the lichens except *F. flaventior*.

The seasonal variation study reports that in two different seasons these lichens do not host the same endolichenic fungal diversity. Out of these four lichen species, *P. reticulatum* shows major differences in endolichenic fungi composition in both seasons.

The endolichenic fungi found ubiquitously in all these macrolichens were: *Alternaria alternata*, *Fusarium oxysporum* and *Trichoderma viride*, while the

Table 2 Endolichenic fungi isolated from some macrolichens of Champawat district

Lichen species	Family	Endolichenic fungi
Canoparmelia texana	Parmeliaceae	*Aureobasidium pullulans*, *Bispora* sp., *Chaetomium* sp., *Dipodascus* (= *Geotrichum*) sp., mycelia sterilia, *Sporormiella minima*, *Penicillium* (= *Torulomyces*) sp., *Xylaria* sp.
Heterodermia diademata	Physciaceae	*A. pullulans*, *Chaetomium* sp., *Geotrichum* sp., *Humicola* sp., *Monilia* sp., mycelia sterilia, *Sordaria* sp., *S. minima*, *Xylaria* sp.
Heterodermia podocarpa	Physciaceae	*A. pullulans*, *Chaetomium* sp., *Humicola* sp., mycelia sterilia, *Periconia* sp., *Sordaria* sp., *Sporormiella minima*, *Torulomyces* sp., *Xylaria* sp.
Leptogium askotense	Collemataceae	*A. pullulans*, *Chaetomium* sp., mycelia sterilia, *N. oryzae*, *Penicillium* sp., *Periconia* sp., *Sordaria* sp., *Talaromyces* sp., *Torulomyces* sp., *Xylaria* sp.
Lobaria kurokawae	Lobariaceae	*Acremonium* sp., *A. pullulans*, *Chaetomium* sp., *Cladosporium* sp., *Geotrichum* sp., *Humicola* sp., *Monilia* sp., mycelia sterilia, *Penicillium* sp., *Periconia* sp., *Sordaria* sp., *Sporormiella minima*, *Talaromyces* sp., *Torulomyces* sp., *Xylaria* sp.
Parmotrema hababianum	Parmeliaceae	*Alternaria* sp., *A. pullulans*, *Chaetomium* sp., *Humicola* sp., mycelia sterilia, *Periconia* sp., *S. minima*, *Torulomyces* sp., *Xylaria* sp.
Parmotrema tinctorum	Parmeliaceae	*Acremonium* sp., *A. pullulans*, *Botrytis* sp., *Chaetomium* sp., *Geotrichum* sp., *Humicola* sp., mycelia sterilia, *Periconia* sp., *Sordaria* sp., *Sporormiella minima*, *Xylaria* sp.
Phaeophyscia hispidula	Physciaceae	*Aspergillus* sp., *A. pullulans*, *Chaetomium* sp., mycelia sterilia, *Periconia* sp., *Sordaria* sp., *Sporormiella minima*, *Torulomyces* sp., *Trichoderma* sp., *Xylaria* sp.
Punctelia rudecta	Parmeliaceae	*Acremonium* sp., *A. pullulans*, *Chaetomium* sp., *Humicola* sp., mycelia sterilia, *Periconia* sp., *Sordaria* sp., *Spegazzinia lobulata*, *Sporormiella minima*, *Torulomyces* sp., *Trichoderma* sp., *Xylaria* sp.
Ramalina conduplicans	Ramalinaceae	*A. pullulans*, *Botrytis* sp., mycelia sterilia, *Penicillium* sp., *S. minima*, *Xylaria* sp.
Usnea sp.	Parmeliaceae	*A. pullulans*, *Humicola* sp., mycelia sterilia, *Penicillium* sp., *Sordaria* sp., *Sporormiella minima*, *Xylaria* sp.

species confined to only a single macrolichen species were: *Aspergillus ochraceous* to *B. setschwanensis* and *Penicillium citrinum* to *F. flaventior*. However, *Mucor* sp. and *Pencillium* sp. were found in all the macrolichens, except *F. flaventior*.

Fungal genera such as *Alternaria*, *Aspergillus*, *Penicillium*, *Fusarium* and *Trichoderma* were the most frequently isolated species while species such as *Acremonium lichenicola*, *Nigrospora oryzae*, *Sordaria fimicola*, *Xyalria hypoxylon* were host specific and were not isolated frequently.

The major fungal endophyte detected in the summer season was *Trichoderma viride* (45 isolates), followed by *Fusarium oxysporum* (32 isolates) and *Alternaria*

Table 3 Diversity of endolichenic fungi isolated from four selected macrolichens (winter season)

Lichens	Endolichenic fungi	No. of isolates	Colonization rate (%)	Relative frequency	Shannon–Weiner biodiversity index (H′)
Bulbothrix setschwanensis	*Alternaria alternata*	05	05	0.714	0.070
	Aspergillus flavus	10	10	1.428	0.078
	Cylindrosporium sp.	01	01	0.142	2.383
	Neocosmospora solani	05	05	0.714	0.070
	Gilmaniella humicola	10	10	1.428	0.078
	Penicillium sp.	15	15	2.142	0.361
	Mycelia sterilia	20	20	2.857	0.687
Flavoparmelia caperata	*Alternaria alternata*	03	03	0.750	0.051
	Aspergillus flavus	20	20	0005	1.618
	Neocosmospora solani	10	10	2.500	0.523
	Mycelia sterilia	27	27	6.750	2.283
Flavopunctelia flaventior	*Alternaria alternata*	05	05	1.000	0.000
	Aspergillus flavus	10	10	2.000	0.300
	Neocosmospora solani	05	05	1.000	0.000
	Trichoderma viride	10	10	2.000	0.300
	Mycelia sterilia	32	32	6.400	2.158
Parmotrema reticulatum	*Acremonium lichenicola*	02	02	0.166	2.015
	Alternaria alternata	05	05	0.416	0.479
	Aspergillus flavus	10	10	0.833	0.020
	Neocosmospora solani	10	10	0.833	0.020
	Nigrospora oryzae	05	05	0.416	0.479
	Mucor racemosus	02	02	0.166	2.015
	Papulospora sp.	02	02	0.166	2.015
	Penicillium sp.	15	15	1.250	0.030
	Pestalotiopsis maculans	01	01	0.083	3.870
	Sordaria fimicola	01	01	0.083	3.870
	Xylaria hypoxylon	02	02	0.166	2.015
	Mycelia Sterilia	25	25	2.083	3.350

Table 4 Diversity of endolichenic fungi isolated from four selected macrolichens (summer season)

Lichens	Endolichenic fungi	No. of isolates	Colonization rate (%)	Relative frequency	Shannon–Weiner biodiversity index (H′)
Bulbothrix setschwanensis	*Alternaria alternata*	06	06	0.66	0.107
	Aspergillus flavus	11	11	1.22	0.024
	Aspergillus niger	10	10	1.11	6.727
	Aspergillus ochraceus	02	02	0.22	1.433
	Fusarium oxysporum	07	07	0.77	0.042
	Mucor sp.	02	02	0.22	1.433
	Penicillium sp.	08	08	0.88	0.010
	Trichoderma viride	05	05	0.55	0.222
	Mycelia sterilia	32	32	3.55	1.004
Flavoparmelia caperata	*Alternaria alternata*	08	08	1.14	0.010
	Aspergillus flavus	15	15	2.14	0.362
	Fusarium oxysporum	05	05	0.71	0.072
	Mucor sp.	05	05	0.71	0.072
	Penicillium sp.	12	12	1.71	0.178
	Trichoderma viride	10	10	1.42	0.076
	Mycelia sterilia	20	20	2.85	0.684
Flavopunctelia flaventior	*Alternaria alternata*	05	05	0.83	0.021
	Aspergillus niger	10	10	1.66	0.160
	Fusarium oxysporum	12	12	2.00	0.300
	Penicillium citrinum	02	02	0.33	0.768
	Trichoderma viride	17	17	2.83	0.675
	Mycelia sterilia	24	24	4.00	1.203

(continued)

Table 4 (continued)

Lichens	Endolichenic fungi	No. of isolates	Colonization rate (%)	Relative frequency	Shannon–Weiner biodiversity index (H′)
Parmotrema reticulatum	*Alternaria alternata*	10	10	1.42	0.076
	Aspergillus niger	12	12	1.71	0.178
	Fusarium oxysporum	08	08	1.14	0.010
	Mucor sp.	07	07	1.00	0.000
	Penicillium sp.	10	10	1.42	0.076
	Trichoderma viride	13	13	1.85	0.236
	Mycelia sterilia	27	27	3.85	1.136

alternata (29 isolates), while the minor ones were *Aspergillus ochraceus* and *Penicillium citrinum* (2 isolates each).

In the present study it was also noticed that members of ascomycota hold the major share, while zygomycota was represented by a single genus *Mucor* in the four tested species of lichens.

To know the effectiveness of different surface sterilization procedures regarding isolation of endolichenic fungi, three protocols were followed (Table 3.1) (Tripathi and Joshi 2015), and the present study suggests that the modified protocol of Suryanarayanan et al. (2005) is the best protocol for isolating endolichenic fungi, because by following this protocol, maximum numbers of isolates of endolichenic fungi were isolated.

It was also noticed that fungal diversity residing inside lichens (i.e. endolichenic fungi) was somewhat similar with endophytes residing inside the host tree from where the lichens were collected. In winters endolichenic fungi isolated from *B. setschwanensis* were found most similar with the host endophytes and shows the similarity value (0.727), however, in summers *F. caperata* having similarity index of 0.80 hosted most similar endophytes as that of endophytes residing in its host.

Surprisingly, authors found that some of the isolated endolichenic fungi (*Alternaria alternata*, *Aspergillus* sp., *Cladosporium herbarum*, *Curvularia australiensis* (= *Bipolaris australiensis*), *Mucor* sp., *Nigrospora* sp., *Penicillium* sp. and *Trichoderma* sp.) have also been reported earlier as saprophytic fungi from dead or decaying lichen thalli (Hawksworth 1979). Besides this, species of some fungal genera (*viz. Acremonium* Link, *Chalara* (Corda) Rabenh., *Cladosporium* Link, *Coniochaeta* (Sacc.) Cooke, *Epicoccum* Link, *Fusarium* Link, *Karsteniomyces* D. Hawksw., *Phoma* Sacc., *Phyllosticta* Pers. and *Sporormiella* Ellis & Everh.) have also been reported as lichenicolous fungi by various workers (Lawrey and Diederich 2016; Joshi et al. 2016). The colonization of these saprophytic and lichenicolous

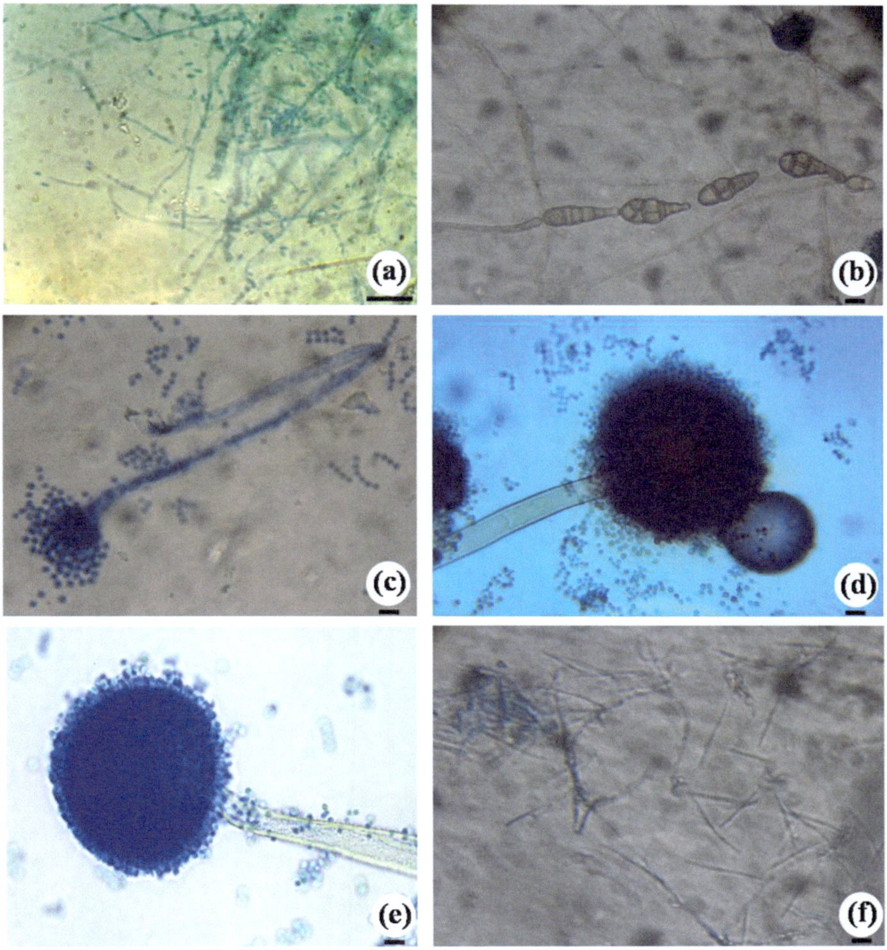

Fig. 1 (**a**) *Acremonium lichenicola*, (**b**) *Alternaria alternata*, (**c**) *Aspergillus flavus*, (**d**) *Aspergillus niger*, (**e**) *Aspergillus ochraceus*, (**f**) *Cylindrosporium* sp. (Scale: 10 μm)

fungal species asymptomatically inside living and healthy lichen thalli is a very interesting finding and provokes us to think that why they are living asymptomatically inside the lichen thallus rather than colonizing on dead/decaying/living thallus and causing visible damage because living asymptomatically is leading them to compromise with their life-style.

Out of 19 endolichenic fungi obtained from 04 macrolichens, 10 were selected and subjected to fermentation for getting secondary metabolites. These fungal strains were cultured on Potato Dextrose Agar at 25 °C for 7 days. Agar plugs were then cut with a cork borer under aseptic conditions and inoculated in three

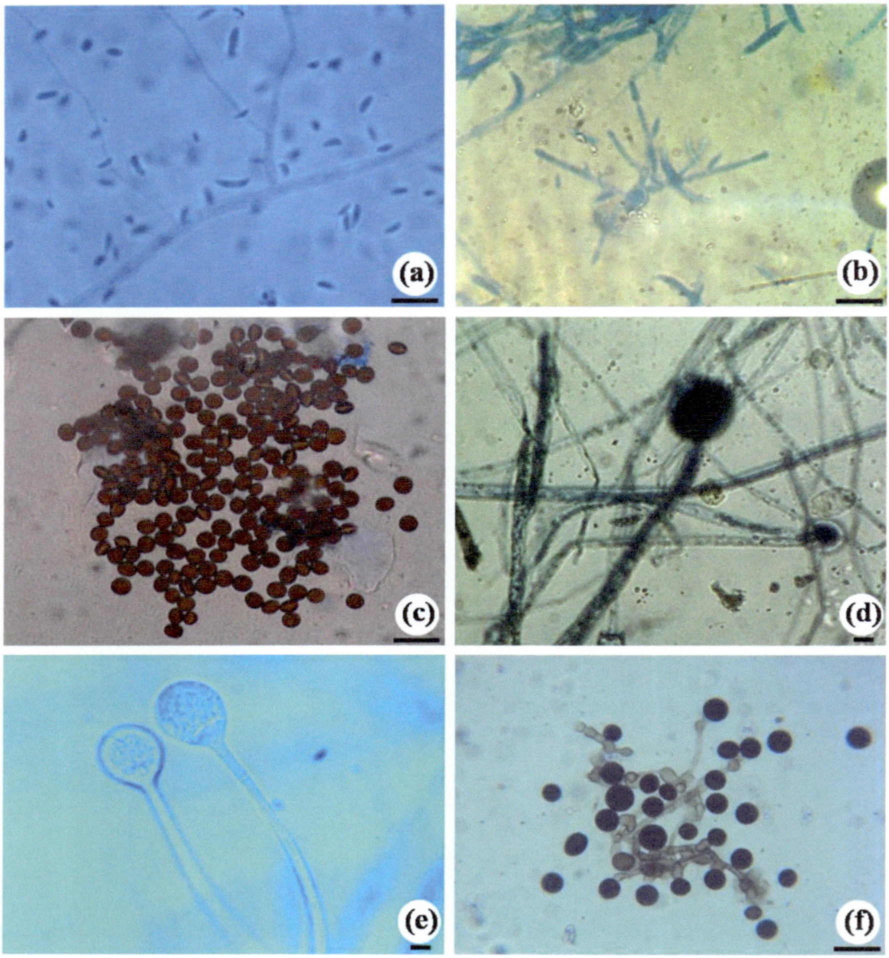

Fig. 2 (**a**) *Fusarium oxysporum*, (**b**) *Neocosmospora solani*, (**c**) *Gilmaniella humicola*, (**d**) *Mucor racemosus*, (**e**) *Mucor* sp., (**f**) *Nigrospora oryzae* (Scale: 10 μm)

Erlenmeyer flasks (250 ml), each containing 50 ml of Potato Dextrose Broth (PDB). These flasks were incubated at 25 °C on a rotary shaker at 170 rpm for 5 days to prepare the spore inoculum.

For the fermentation of fungi a very economically affordable method was chosen i.e. they were cultured in rice media. Fermentation was carried out in 12 Erlenmeyer flasks (500 ml), each containing 80 g of rice. Distilled water (120 ml) was added to each flask, and the contents were soaked overnight before autoclaving. After cooling to room temperature, each flask was inoculated with 5 ml of the spore inoculum and incubated at 25 °C for 40 days (Zhang et al. 2014).

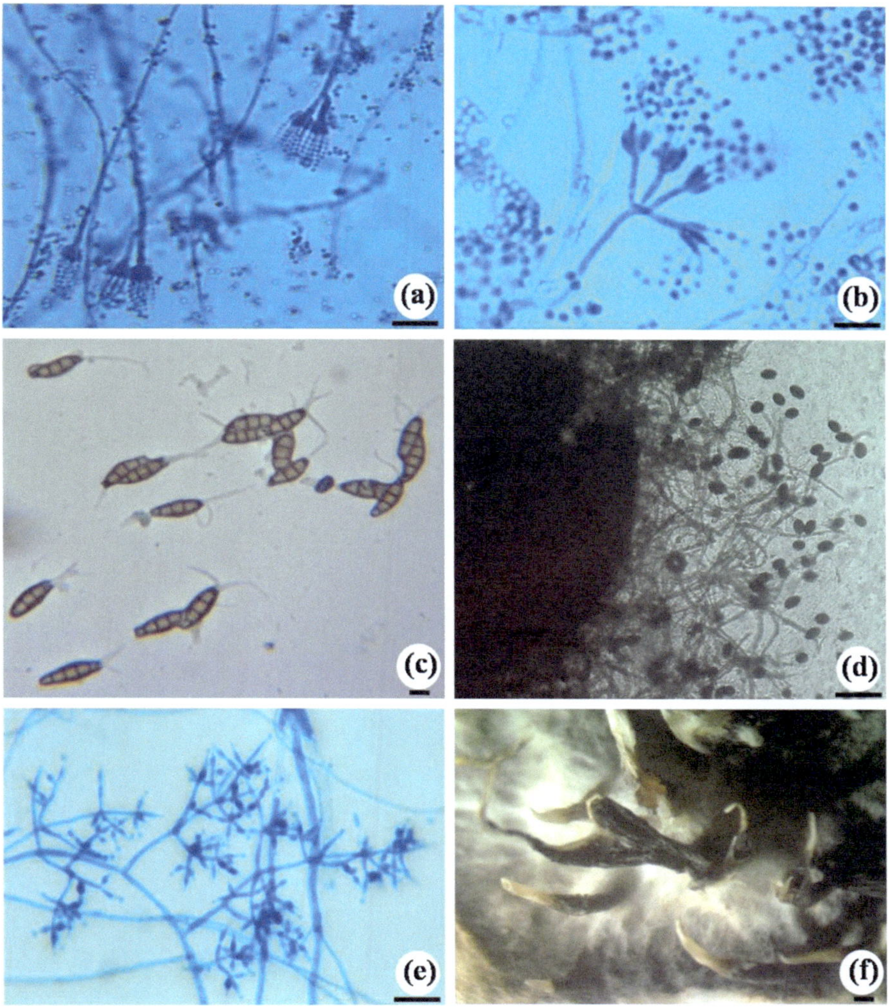

Fig. 3 (**a**) *Penicillium* sp., (**b**) *Penicillium citrinum*, (**c**) *Pestalotiopsis maculans*, (**d**) *Sordaria fimicola*, (**e**) *Trichoderma viride*, (**f**) *Xylaria hypoxylon* (Scale: a–e = 10 μm; f = 1 cm)

The culture was extracted with ethyl acetate. Each Erlenmeyer flask was filled with 100 ml ethyl acetate and then left overnight for proper isolation of secondary metabolites and next day this 100 ml of solvent was transferred into another flask and at each and every transfer 50 ml fresh ethyl acetate was added to the new flask to avoid saturation of solvent by secondary metabolites present in the culture. After that the organic solvent was evaporated to dryness to have the crude extract.

After having the crude extract it was tested against a panel of crop and human pathogenic bacteria. The activity profile of secondary metabolites from endoli-

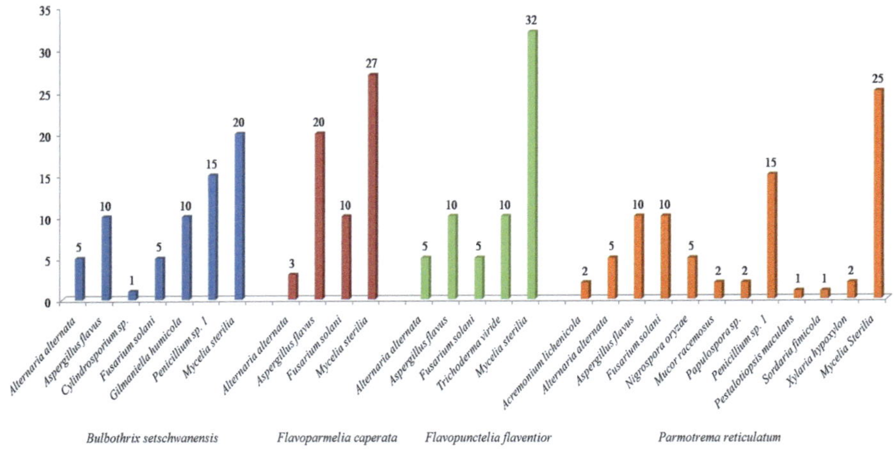

Fig. 4 Number of fungal colonies isolated from 4 macrolichens in winter season

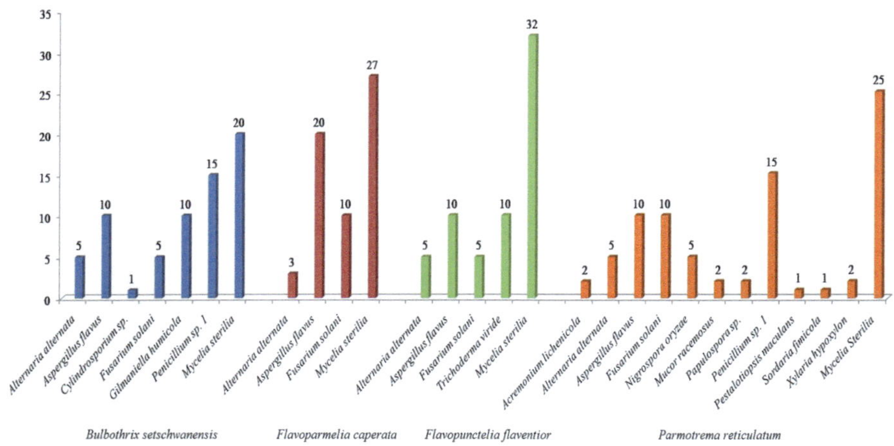

Fig. 5 Number of fungal colonies isolated from 4 macrolichens in summer season

chenic fungi against the tested bacteria is shown in Tables 5 and 6. Results showed that crude extracts of endolichenic fungi exhibited inhibitory activity against more or less all the tested bacteria (Tables 5 and 6).

Among all the tested crop pathogens *X. campestris* was found to be the most susceptible against 7 out of 10 tested fungal extracts and showed zone of inhibition more than 10 mm in all 7 fungal extracts. However, *R. solanacearum* was the most

Table 5 Antibacterial activity ethyl acetate extract of endolichenic fungi against crop pathogenic bacterial strains

S. No.	Endolichenic fungi	Crop pathogenic bacterial strains *Xanthomonas oryzae* ZOI	MIC	*Xanthomonas campestris* ZOI	MIC	*Xanthomonas phaseoli* ZOI	MIC	*Ralstonia solanacearum* ZOI	MIC	*Erwinia chrysanthemi* ZOI	MIC	*Agrobacterium tumefaciens* ZOI	MIC
1	A01	6.33 ± 0.57	300	11.00 ± 1.00	200	7.33 ± 1.52	200	6.33 ± 0.57	300	7.00 ± 1.00	300	8.00 ± 1.00	200
2	A05	9.66 ± 0.57	300	10.66 ± 0.57	200	6.66 ± 1.15	300	7.66 ± 0.57	300	7.66 ± 0.57	300	7.00 ± 1.00	300
3	A11	9.33 ± 0.57	200	17.66 ± 1.52	200	6.33 ± 0.57	300	6.66 ± 0.57	200	6.33 ± 0.57	300	11.33 ± 0.57	200
4	B01	15.33 ± 1.15	200	10.33 ± 0.57	200	9.33 ± 0.57	200	9.66 ± 0.57	200	10.66 ± 0.57	200	11.00 ± 1.00	200
5	B03	9.33 ± 0.57	200	9.66 ± 0.57	200	11.66 ± 0.57	200	6.33 ± 0.57	200	13.33 ± 0.57	200	13.00 ± 1.00	200
6	D02	6.33 ± 0.57	200	11.33 ± 0.57	200	7.66 ± 0.57	200	6.33 ± 0.57	300	10.33 ± 0.57	200	8.33 ± 0.57	200
7	D31	9.33 ± 0.57	200	9.66 ± 0.57	200	7.33 ± 0.57	300	6.66 ± 0.57	300	7.33 ± 0.57	300	9.33 ± 0.57	300
8	E03	11.33 ± 0.57	300	16.00 ± 1.73	200	11.33 ± 0.57	200	6.33 ± 0.57	300	12.66 ± 0.57	200	6.33 ± 0.57	300
9	E27	10.33 ± 0.57	300	10.33 ± 0.57	300	8.33 ± 0.57	300	7.33 ± 0.57	300	8.66 ± 0.57	300	10.33 ± 0.57	200
10	E48	9.00 ± 1.00	300	9.66 ± 1.15	300	7.33 ± 0.57	300	6.33 ± 0.57	300	7.66 ± 1.15	300	7.66 ± 1.15	300
11	Streptomycin (reference antibiotic)	24.33 ± 1.52	100	23.66 ± 1.52	100	32.66 ± 2.08	50	25.66 ± 0.57	100	25.66 ± 0.57	50	33.00 ± 1.00	50

A01- Mycelia sterilia, A05- *Penicillium* sp., A11- *Mucor* sp., B01- *Trichoderma viride*, B03-*Penicillium citrinum*, D02- *Penicillium* sp., D31- Mycelia sterilia, E03- *Aspergillus ochraceous*, E27- *Aspergillus niger*, E48-Mycelia sterilia

Table 6 Antibacterial activity ethyl acetate extract of endolichenic fungi against human pathogenic bacterial strains

		Human pathogenic bacterial strains									
		Escherichia coli		*Bacillus subtilis*		*Pseudomonas aeruginosa*		*Salmonella typhimurium*		*Klebsiella pneumoniae*	
S. No.	Endolichenic fungi	ZOI	MIC	ZOI	MIC	ZOI	MIC	ZOI	MIC	ZOI	MIC
1	A01	7.33 ± 1.52	300	7.33 ± 1.52	300	7.00 ± 1.73	300	6.33 ± 0.57	300	9.00 ± 1.73	200
2	A05	7.33 ± 1.52	200	7.00 ± 1.00	200	7.00 ± 1.00	300	12.00 ± 1.00	200	11.66 ± 0.57	200
3	A11	11.33 ± 0.57	200	10.66 ± 0.57	200	8.33 ± 1.73	200	9.00 ± 0.57	200	12.66 ± 0.57	200
4	B01	15.33 ± 1.15	200	14.33 ± 2.08	200	10.00 ± 1.73	200	12.66 ± 1.15	200	12.00 ± 1.00	200
5	B03	13.33 ± 0.57	200	16.66 ± 0.57	200	6.33 ± 0.57	200	11.66 ± 0.57	200	9.33 ± 0.57	200
6	D02	12.66 ± 0.57	200	7.33 ± 0.57	200	6.33 ± 0.57	300	9.66 ± 0.57	300	8.33 ± 0.57	200
7	D31	8.66 ± 1.15	300	6.33 ± 0.57	300	6.33 ± 0.57	300	11.33 ± 1.52	200	10.00 ± 1.00	200
8	E03	11.66 ± 0.57	200	6.33 ± 0.57	300	6.33 ± 0.57	300	10.66 ± 0.57	200	9.66 ± 0.57	200
9	E27	10.33 ± 0.57	200	7.66 ± 0.57	200	7.33 ± 0.57	300	10.33 ± 0.57	200	7.33 ± 0.57	300
10	E48	8.66 ± 1.15	200	6.33 ± 0.57	200	6.33 ± 0.57	300	6.66 ± 0.57	300	7.66 ± 1.15	300
11	Streptomycin (reference antibiotic)	31.66 ± 1.15	50	26.33 ± 0.57	100	27.33 ± 1.15	100	24.66 ± 1.15	100	30.33 ± 0.57	100

A01- Mycelia sterilia, A05- *Penicillium* sp., A11- *Mucor* sp., B01- *Trichoderma viride*, B03- *Penicillium citrinum*, D02- *Penicillium* sp., D31- Mycelia sterilia, E03- *Aspergillus ochraceous*, E27- *Aspergillus niger*, E48-Mycelia sterilia

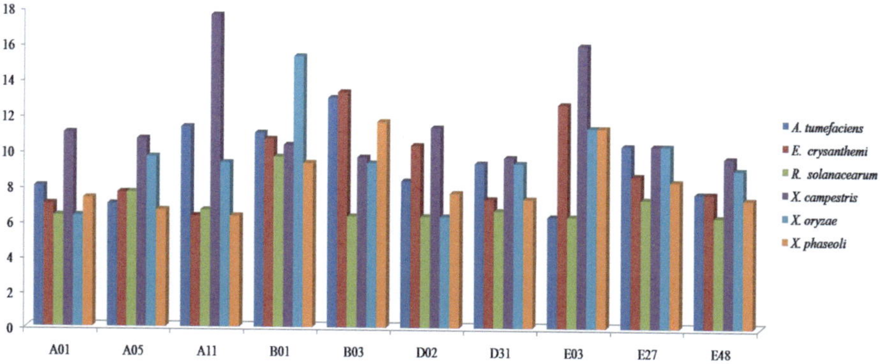

Fig. 6 Antimicrobial activity of endolichenic fungi against crop pathogens. (A01- Mycelia sterilia, A05- *Penicillium* sp., A11- *Mucor* sp., B01- *Trichoderma viride*, B03- *Penicillium citrinum*, D02- *Penicillium* sp., D31- Mycelia sterilia, E03- *Aspergillus ochraceous*, E27- *Aspergillus niger*, E48-Mycelia sterilia)

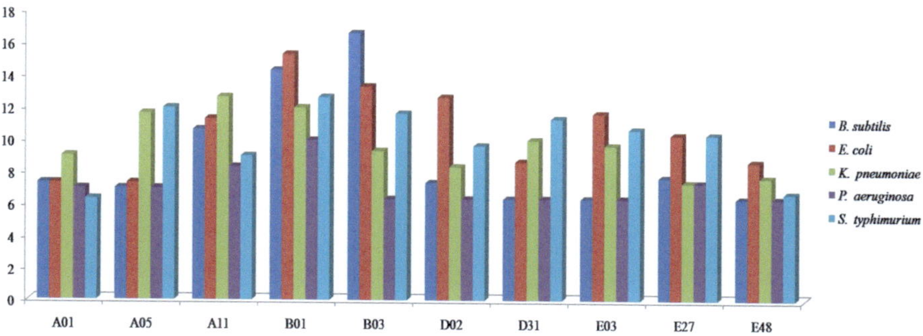

Fig. 7 Antimicrobial activity of endolichenic fungi against human pathogens (A01- Mycelia sterilia, A05- *Penicillium* sp., A11- *Mucor* sp., B01- *Trichoderma viride*, B03- *Penicillium citrinum*, D02- *Penicillium* sp., D31- Mycelia sterilia, E03- *Aspergillus ochraceous*, E27- *Aspergillus niger*, E48-Mycelia sterilia)

resistant pathogen since it showed zone of inhibition less than 10 mm against all the tested fungal extracts (Fig. 6).

Among all the tested human pathogens *E. coli* and *S. typhimurium* were found to be equally susceptible against 6 out of 10 fungal extracts and showed zone of inhibition more than 10 mm, while *P. aeruginosa* was the most resistant pathogen since it showed zone of inhibition less than 10 mm against 9 out of 10 tested fungal extracts (Fig. 7).

Among 11 tested pathogens, extract of *Aspergillus niger* came out as a broad spectrum antibiotic since it showed maximum zone of inhibition against 05 pathogens (*X. oryzae*, *X. campestris*, *A. tumefaciens*, *E. coli* and *S. typhimurium*) and showed minimum inhibition against *E. chrysanthemi* (8.66 mm; MIC = 300 mg/ml).

Table 7 Major chemical constituents of *Mucor* sp.

S. No.	Retention time	Area %	Name of compound
1	13.417	1.06	8-Methylenebicyclo[2.2.2]oct-5-en-2-one
2	20.420	1.25	1-Tetradecene
3	20.636	3.86	Tetradecane
4	23.676	3.67	Phenol, 2,4-bis(1,1-dimethylethyl)-
5	24.879	1.16	Pentadecane, 3-methyl-
6	25.410	2.51	1-Heptadecene
7	**25.600**	**5.01**	**Hexadecane**
8	25.724	2.95	Pyridine, 2-methyl-5-(1-methylethenyl)-
9	26.019	2.49	2-Pyridinecarboxylic acid, 5-butyl-
10	29.875	2.24	1-Heptadecene
11	30.026	3.98	Octadecane
12	32.621	2.21	Hexadecanoic acid, methyl ester
13	**33.496**	**5.56**	**Octadecanoic acid**
14	33.924	1.32	1-Heneicosanol
15	34.047	2.16	Eicosane
16	35.878	1.12	9,12-Octadecadienoic acid, methyl ester
17	**36.017**	**8.75**	**9-Octadecenoic acid, methyl ester, (E)-**
18	36.944	2.92	Bis(2-ethylhexyl) maleate
19	37.187	2.76	Octadecanoic acid
20	37.187	1.54	Hexadecanoic acid, butyl ester
21	40.656	1.35	Cholest-22-ene-21-ol, 3,5-dehydro-6-methoxy-, pivalate
22	**43.847**	**10.40**	**6-Hydroxy-7-isopropyl-1,4a-dimethyl-1,2,3,4,4a,9,10,10a-**
23	44.199	10.92	Unidentified

Three strains of Mycelia sterilia (A01, D31 and E48) were also screened for their antimicrobial activity and they showed moderate antimicrobial activity. One of them (D31) was found inhibiting two human pathogens *viz. S. typhimurium* and *K. pneumoniae* with zones of inhibition 11.33 and 10.00 mm, respectively, and A01 was found inhibiting *X. campestris* with zone of inhibition 11.00 mm.

Out of the 10 endolichenic fungi which were selected to evaluate their bioactive potentials, only 4 were studied in detail to know their chemical composition and bio-prospection values, since these 4 fungal species were the only ones which were fast growing and yielding high amount of secondary metabolites. The fungal metabolites extracted from these 4 species were subjected to GC-MS analysis and all of them displayed a plethora of compounds (Tables 7, 8, 9, and 10).

Mucor sp. showed the predominance of Hexadecane; Octadecanoic acid; 8(E)-Octadecenoic acid, methyl ester; 6-Hydroxy-7-isopropyl-1,4a-dimethyl-1,2,3,4,4a,9,10,10a- (Figs. 8, 9, and 10).

Table 8 Major chemical constituents of *Trichoderma viride*

S. No.	Retention time	Area %	Name of compound
1	20.417	1.09	1-Tetradecene
2	20.633	3.37	Tetradecane
3	23.677	3.30	Phenol, 2,4-bis(1,1-dimethylethyl)-
4	25.401	1.99	1-Heptadecene
5	25.584	4.38	Hexadecane
6	29.871	1.72	1-Heptadecene
7	32.616	1.54	Hexadecanoic acid, methyl ester
8	33.477	3.66	n-Hexadecanoic acid
9	34.043	1.85	Eicosane
10	**36.007**	**6.78**	**9-Octadecenoic acid, methyl ester, (E)-**
11	36.947	2.56	Bis(2-ethylhexyl) maleate
12	37.512	1.51	Hexadecanoic acid, butyl ester
13	38.341	4.06	1-Hydroxy-4-methylanthraquinone
14	40.398	1.17	9,10-Anthracenedione, 1,8-dihydroxy-3-met
15	**42.302**	**5.32**	**Simvastatin, trimethylsilyl ether**
16	43.331	1.97	2-Butenoic acid, 2-methyl-, dodecahydro-8-hydroxy-8a-m
17	**44.003**	**11.14**	**Cylobutanecarboxylicacid,2,6-dimethylnon-1-en-3-yn-5-**
18	45.211	2.03	Butane-1,1-dicarbonitrile, 1-cyclohexyl-3-methyl-
19	45.358	2.24	Cyclopropanecarboxylicacid, 1-methyl-, 2,6-bis(1,1-dimet
20	**47.177**	**11.06**	**Docosa-8,14-diyn-cis-1,22-diol, bis(trimethylsilyl) ether**
21	48.747	1.26	Propanoic acid, 2-methyl-, (decahydro-6a

Table 9 Major chemical constituents of *Penicillium citrinum*

S. No.	Retention time	Area %	Name of compound
1	4.823	1.01	2-Hydroxy-2-methyl-4-pentanone(Diacetone
2	19.256	1.34	Phenol, 2,4-bis(1,1-dimethylethyl)-
3	**28.112**	**4.92**	**9-Octadecenoic acid (Z)-, methyl ester**
4	**28.153**	**1.49**	**Anthralin**
5	28.652	1.26	Cis-Vaccenic acid
6	28.787	1.23	Bis(2-ethylhexyl) maleate
7	33.160	1.47	9,10-Anthracenedione, 1,8-Dihydroxy-3-metho
8	**34.816**	**33.89**	**Spiro[benzofuran-2(3H),1'-[2] cyclohexene]-3,4'-dione,2',4,6**
9	**36.627**	**37.66**	**Spiro[benzofuran-2(3H),1'-[2] cyclohexene]-3,4'**

Table 10 Major chemical constituents of *Aspergillus niger*

S. No.	Retention time	Area %	Name of compound
1	22.938	1.64	3-Furanacetic acid, 4-hexyl-2,5-dihydro-2,5-dioxo-
2	23.670	3.65	Phenol, 3,5-bis(1,1-dimethylethyl)-
3	24.878	1.22	Tridecane, 1-iodo-
4	25.393	1.71	1-Pentadecene
5	25.569	4.21	Hexadecane
6	29.873	3.16	1-Heptadecene
7	**30.019**	**5.95**	**Octadecane**
8	30.070	1.49	Cyclohexanone, 2-(phenylmethylene)-
9	32.615	3.09	Hexadecanoic acid, methyl ester
10	**33.456**	**7.45**	**n-Hexadecanoic acid**
11	33.919	1.75	1-Heneicosanol
12	34.040	3.14	Eicosane
13	35.875	1.82	9,12-Octadecadienoic acid (Z,Z)-, methyl ester
14	**36.004**	**14.56**	**9-Octadecenoic acid, methyl ester, (E)-**
15	36.655	1.24	9,12-Octadecadienoic acid (Z,Z)-
16	36.938	4.16	Bis(2-ethylhexyl) maleate
17	37.158	1.69	Octadecanoic acid
18	37.502	3.17	Hexadecanoic acid, butyl ester
19	37.714	1.47	Heneicosane
20	38.496	1.58	Bicyclo[2.2.1]heptane, 2-(phenylmethyl)-
21	**43.177**	**14.93**	**Phenol, 3,5-dimethoxy-, acetate**

Trichoderma viride signifies the major occurrence of 9-Octadecenoic acid, methyl ester, (E)-; Simvastatin, trimethylsilyl ether; Cylobutanecarboxylic acid, 2,6-dimethylnon-1-en-3-yn-5-; Docosa-8,14-diyn-cis-1,22-diol, bis(trimethylsilyl) ether (Figs. 11, 12, and 13).

Penicillium citrinum exhibit the key existence of Spiro[benzofuran-2(3H),1'-[2] cyclohexene]-3,4'-dione,2',4,6; Spiro[benzofuran-2(3H),1'-[2] cyclohexene]-3,4'; 9-Octadecenoic acid (Z)-, methyl ester; Anthralin (Figs. 14, 15, and 16).

Aspergillus niger contains Octadecane; n-Hexadecanoic acid; 9-Octadecenoic acid, methyl ester, (E)-; Phenol, 3,5-dimethoxy-, acetate as chief chemical constituents (Figs. 17, 18, and 19).

In nut shell the present study unveils the hidden diversity of endolichenic fungi and put forward a natural, untapped and inexhaustible alternative to the presently available drugs in markets for both crops as well as humans. These fungi are understudied and less explored groups of microbial symbiont. As mentioned earlier, surveys in recent past suggested that they are the producers of effective bioactive

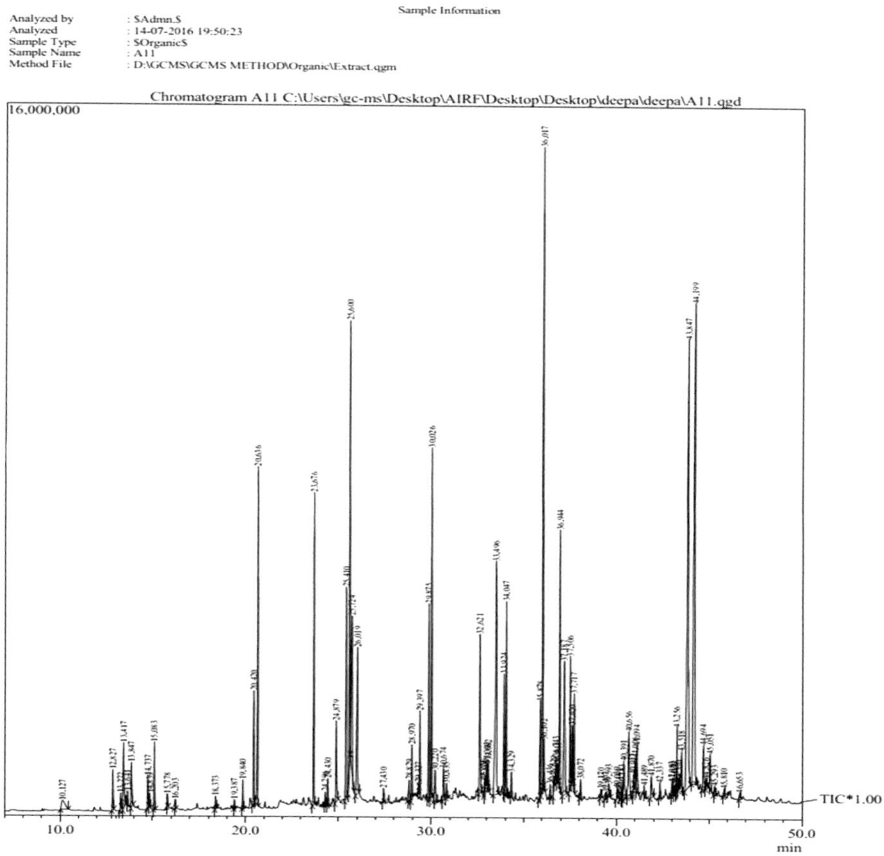

Fig. 8 GCMS report of *Mucor* sp.

metabolites with wide therapeutic applications, hence, considering the diminishing lichen diversity and slow growing nature, the research priority should be directed to study these fungi in India, because they are capable of producing a vast array of compounds which are either similar to lichens or are novel. The current study is an endeavor in this direction, and suggests that endolichenic fungi could be a potential source of antimicrobial agents.

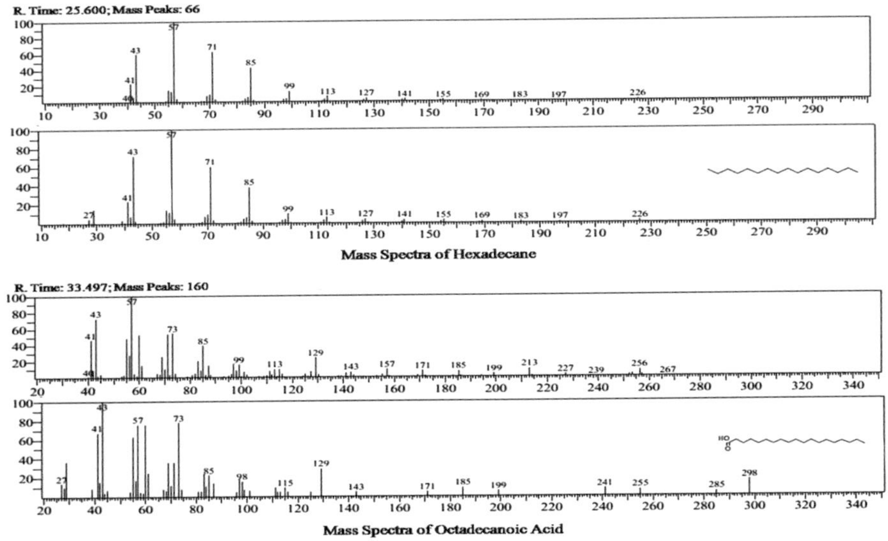

Fig. 9 Mass spectra of major compounds extracted from *Mucor* sp.

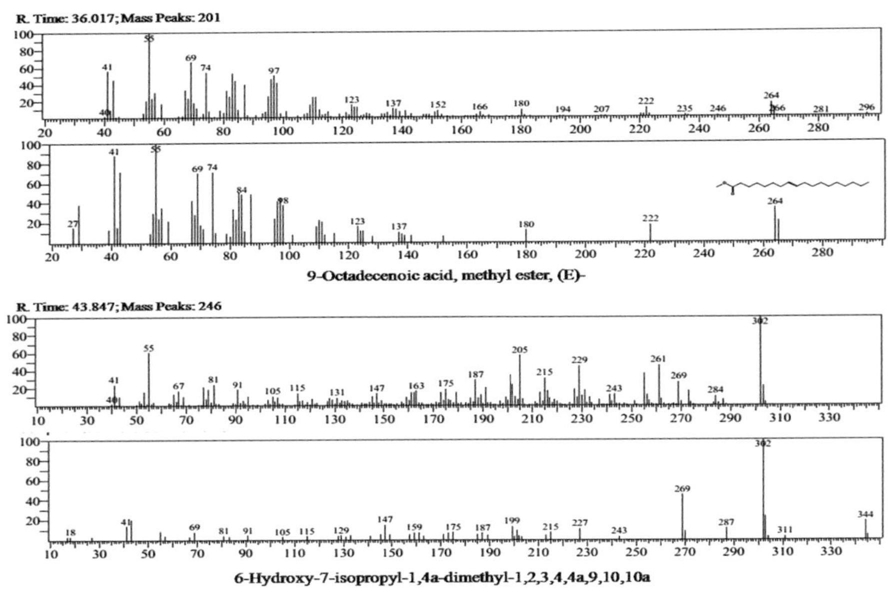

Fig. 10 Mass spectra of major compounds extracted from *Mucor* sp.

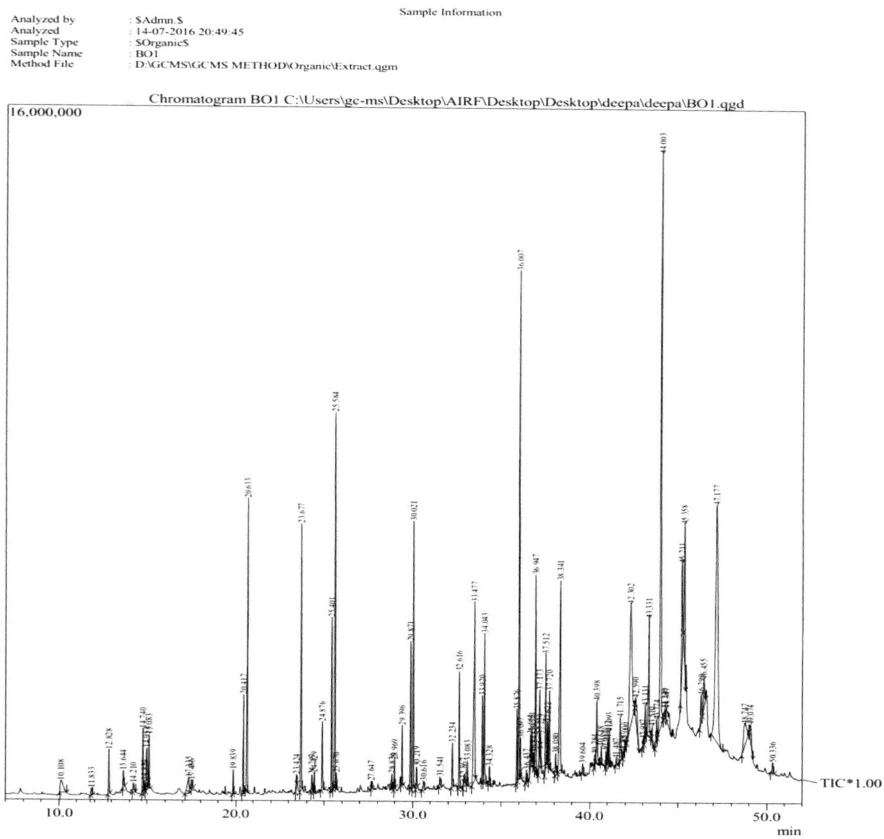

Fig. 11 GCMS report of *Trichoderma viride*

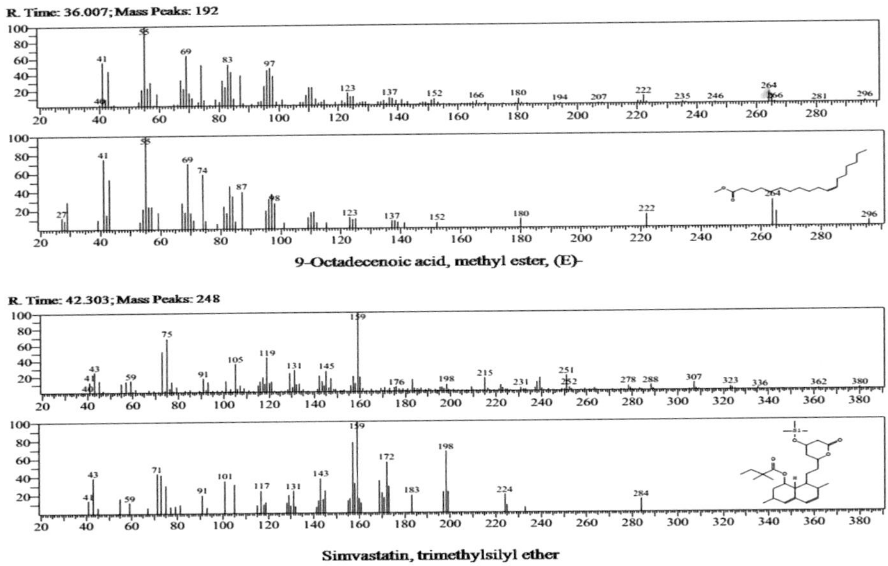

Fig. 12 Mass spectra of major compounds extracted from *Trichoderma viride*

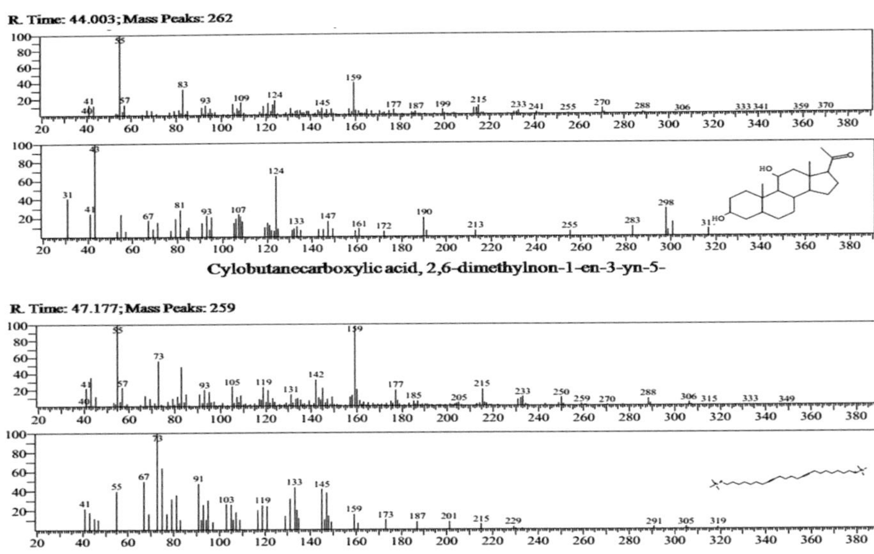

Fig. 13 Mass spectra of major compounds extracted from *Trichoderma viride*

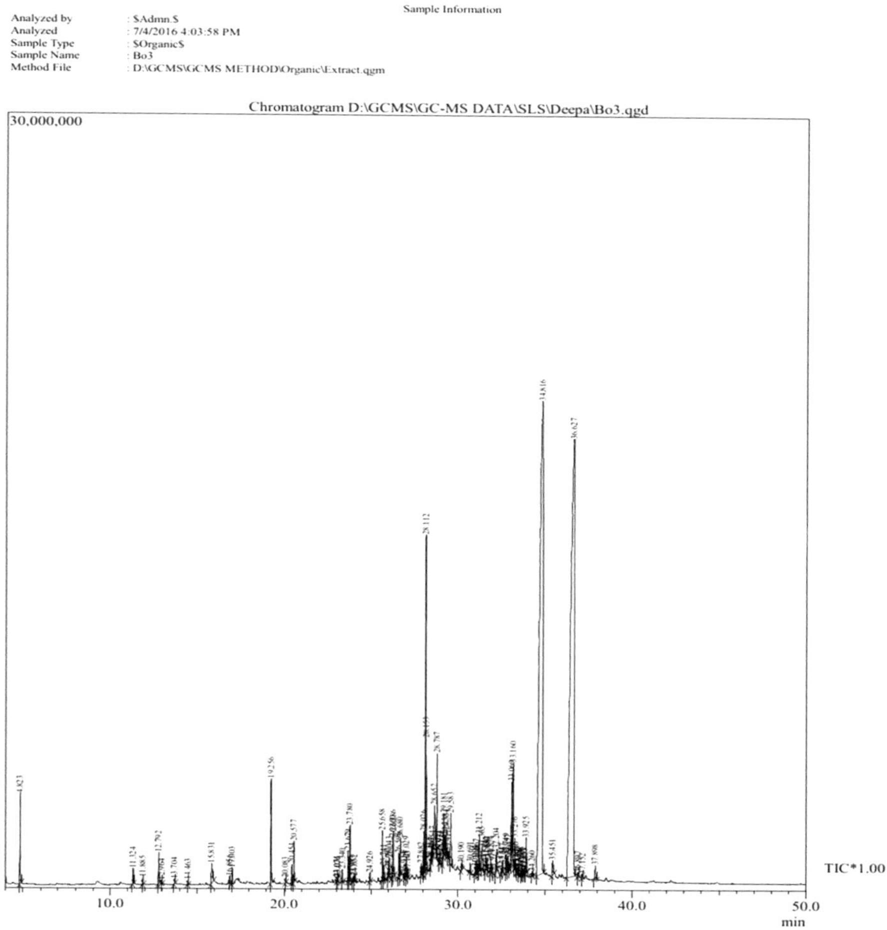

Fig. 14 GCMS report of *Penicillium citrinum*

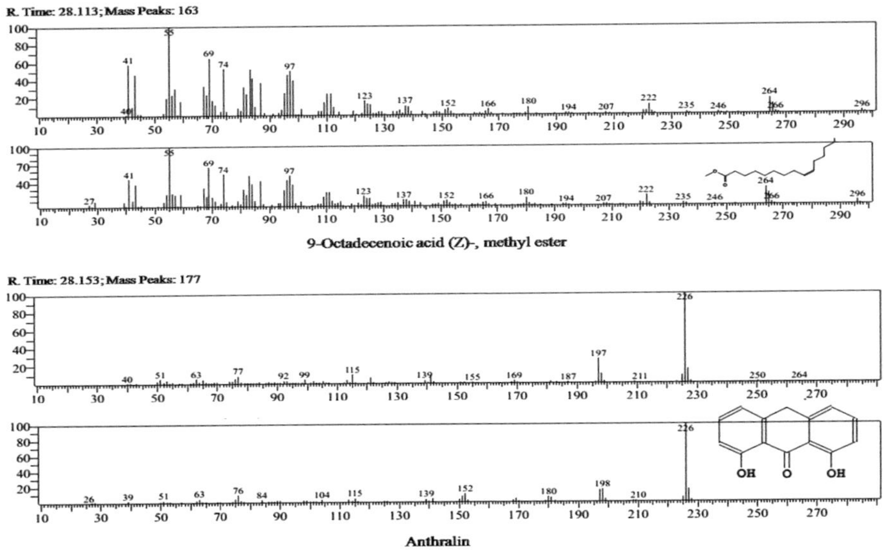

Fig. 15 Mass spectra of major compounds extracted from *Penicillium citrinum*

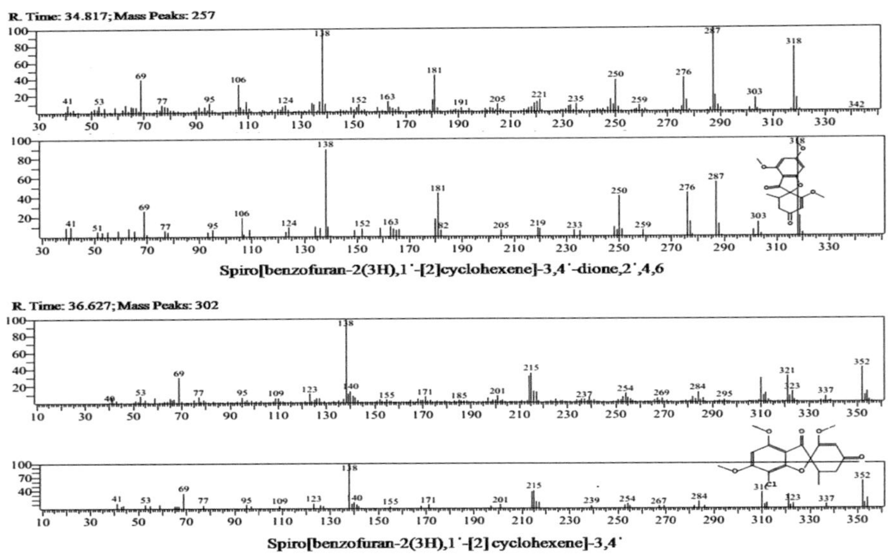

Fig. 16 Mass spectra of major compounds extracted from *Penicillium citrinum*

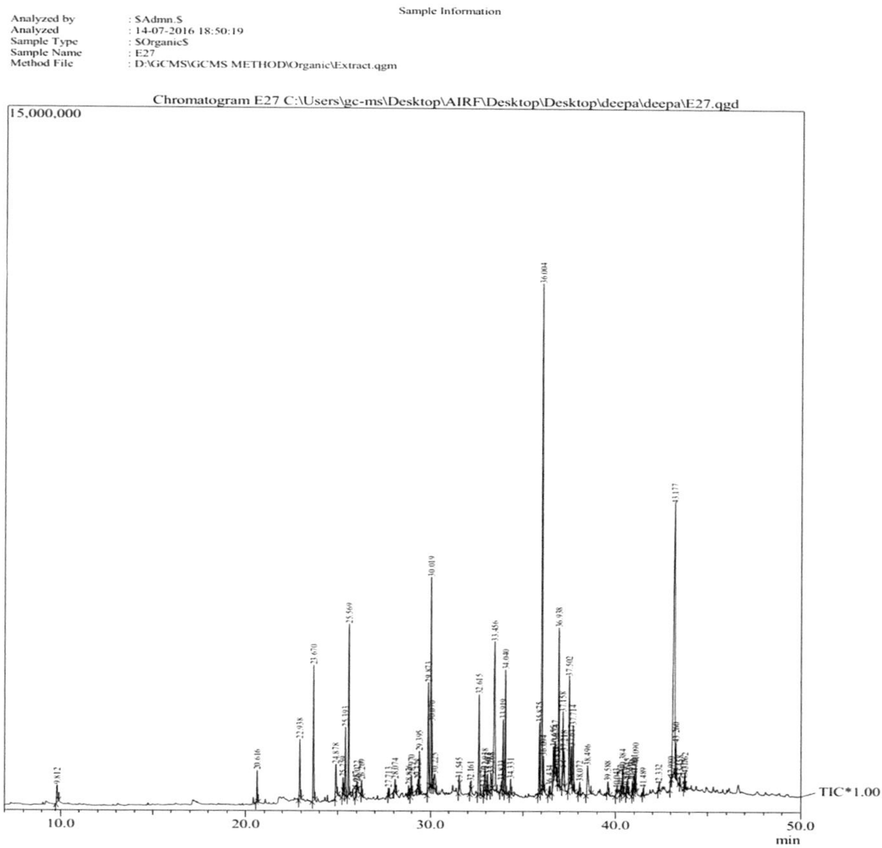

Fig. 17 GCMS report of *Aspergillus niger*

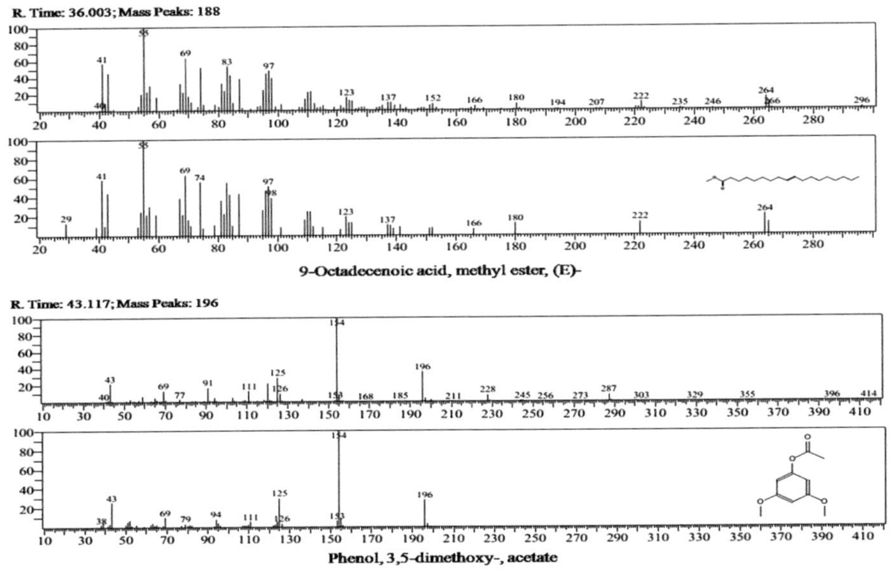

Fig. 18 Mass spectra of major compounds extracted from *Aspergillus niger*

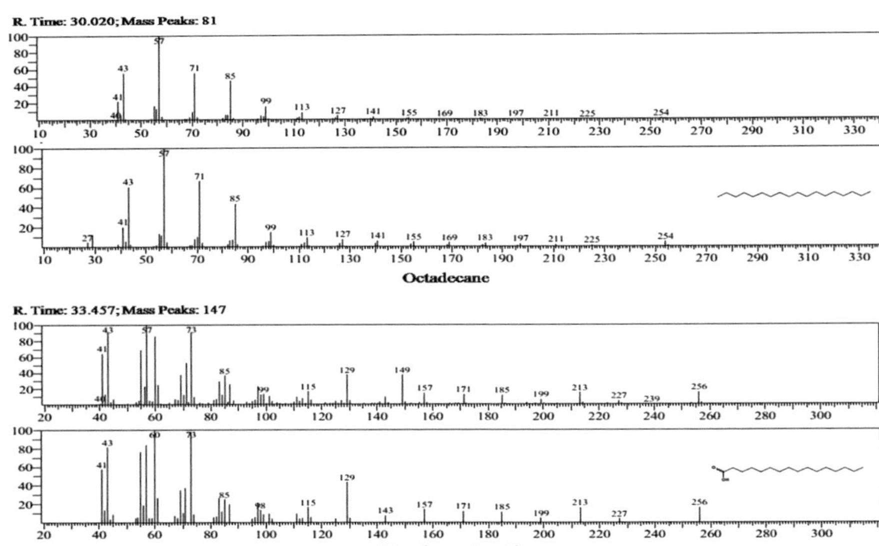

Fig. 19 Mass spectra of major compounds extracted from *Aspergillus niger*

References

Awasthi DD (2007) A compendium of the Macrolichens from India, Nepal and Sri Lanka. Bishen Singh Mahendra Pal Singh Publication, Dehra Dun

Chowdhry PN (2000) Manual on identification of plant pathogenic and bio-control fungi of agricultural importance. IARI, New Delhi

Ellis MB (1971) Dematiaceous hyphomycetes. CABI Publications, Commonwealth Mycological Institute, Wallingford/Kew, pp 608

Ellis MB (1976) More dematiaceous hyphomycetes. CABI Publications, Commonwealth Mycological Institute, Wallingford/Kew, pp 507

Gilman JC (1967) Manual of soil fungi. Oxford/IBH Publishing, New Delhi, pp 450

Harris JL (2000) Safe, low-distortion tape touch method for fungal slide mounts. J Clin Microbiol 38:4683–4684

Hawksworth D (1979) The lichenicolous Hyphomycetes. Bull Brit Mus (Nat Hist) Bot 6:183–300

Joshi Y, Falswal A, Tripathi M et al (2016) One hundred and five species of lichenicolous biota from India: an updated checklist for the country. Mycosphere 7(3):268–294

Lawrey JD, Diederich P (2016) Lichenicolous fungi – worldwide checklist, including isolated cultures and sequences available. http://www.lichenicolous.net

Schulz B, Wanke U, Draeger S et al (1993) Endophytes from herbaceous plants and shrubs-effectiveness of surface sterilization methods. Mycol Res 97:1447–1450

Sinha GP, Nayaka S, Joseph S (2018) Additions to the checklist of Indian lichens after 2010. Cryptogam biodiversity and assessment special volume 197–206

Suryanarayanan TS, Thirunavukkarasu N, Hariharan GN et al (2005) Occurrence of non-obligate microfungi inside lichen thalli. Sydowia 57(1):20–130

Sutton BC (1980) The Coelomycetes: fungi imperfecti with pycnidia, acervuli and stromata. CABI Publications, Commonwealth Mycological Institute, Wallingford/Kew, pp 696

Tripathi M, Joshi Y (2015) Endolichenic fungi in Kumaun Himalaya: a case study. In: Upreti DK, Divakar PK, Shukla V et al (eds) Recent advances in lichenology. Springer, New Delhi, pp 111–120

Tripathi M, Gupta RC, Joshi Y (2014a) *Spegazzinia tessarthra* isolated as a true endophyte from lichen *Heterodermia flabellata*. Indian Phytopathol 67(1):109–110

Tripathi M, Gupta RC, Joshi Y (2014b) *Physcia dilatata* Nyl. (lichenized fungi, Physciaceae); a new host of *Bipolaris australiensis* (M.B. Ellis) Tsuda and Ueyama from Kumaun Himalaya, India. Proc Nat Acad Sci Lett 37(5):477–479

Tripathi M, Joshi Y, Gupta RC (2014c) Assessment of endolichenic fungal diversity in some forests of Kumaun Himalaya. Curr Sci 107(5):745–748

Zhang K, Ren J, Ge M et al (2014) Mono- and bis-furanone derivatives from the endolichenic fungus *Peziza* sp. Fitoterapia 92:79–84

Taxonomic Descriptions of Endolichenic Fungi

Abstract Each identified fungal genera and species is given a taxonomic description in the mycological terminology at the time of its publication which is used as identification guidelines in the further studies. So the taxonomic description is the universal identification scripture written specifically about that particular species and this exhibits its significance. Hence, this chapter entails the brief descriptions of the endolichenic fungal genera identified till date. This is the first attempt of its kind in which the authors have compiled the taxonomic descriptions of all the endolichenic fungal genera known so far across the globe.

Talbot (1971) rightly said that taxonomy is largely concerned with determining species and if the species is defined in terms upon which everyone should agree than we will face less difficulty in determining them. Since fungi vary a lot, hence, species in them are essentially unstable and are defined in various ways according to one's interest in different aspects of their variation. Some people are interested in the morphological, others in the genetic variation of fungi, and each has different ideas on what constitutes a species (Talbot 1971).

A great deal of difficulty is often being faced by a person while identifying culture fungi rather than specimens taken from nature, as fruit bodies formed in culture may sometimes differ greatly from those of the same species in nature, for example, synnemata of natural fructifications may become sporodochia in culture; sporodochia may become discrete sporophores; and above all the culture medium cannot tell us whether pycnidia are immersed or erumpent on natural substrata. Besides this, the type specimens, descriptions, keys and whole system of classification which are mainly based on natural fruit bodies and thallus organization, get collapsed with some fungi in culture. Mason (1940) while studying culture fungi stated that "*their study is not the study of plants as they occur, but the study of plants as they do not occur. The game is up if they are not named before they are planted in a test tube*". Again in 1948, he stated that "*their form [morphology] need not be established in order, so that they can be taxonomised into species; it is their pedigree which must be established*". These contradictory statements clearly show the clash between species based on phenotypes and on genotypes. Mason also pointed

M. Tripathi, Y. Joshi, *Endolichenic Fungi: Present and Future Trends*,
https://doi.org/10.1007/978-981-13-7268-1_7

out that one tends to study individual organs and not an individual fungus in all its phases, and that one cannot establish the full range of variation of a species simply by taking a single isolate and multiplying it repeatedly in culture. Thus, species are fairly arbitrary units and the main aim of taxonomist is to identify and name them.

Though ample literature is available pertaining to description of various species of fungi, lichens and lichenicolous fungi, but till date not a single literature is available describing an endolichenic fungus either in brief or in detail. Hence, this led us to frame out this chapter, describing in brief about the so far known 70 genera of endolichenic fungi known across the world by consulting various literatures *viz.* Ando et al. (1998), Arenal et al. (2005), Arx (1981), Barnett (1962), Barnett & Hunter (1998), Bissett (1983), Dodge (1937), Ellis (1971, 1976), Fallah et al. (1998), Hawksworth (1980, 1981), Hirooka et al. (2012), Khan & Cain (1979), Kolařík et al. (2017), Nag Raj (1993), Rahnama and Habibi (2015), Rodrigues et al. (1993), Subramanian (1983), Sutton (1980), Umabala et al. (2001), Yuan and Zhao (1992) along with internet source. The genera are cited as per MycoBank and Index Fungorum.

1. ***Acremonium*** Link, Magazin der Gesellschaft Naturforschenden Freunde Berlin 3 (1): 15 (1809)

Diagnostic characters Colonies are usually slow growing, often compact and moist at first, becoming powdery, suede-like or floccose with age, and may be white, gray, pink, rose or orange in color. Mycelium prostrate, slender. Hyphae often crustaceous, fine, hyaline and produce mostly simple awl-shaped erect phialides. Conidiophores erect, simple, septate, hyaline, gradually tapering toward apex with terminal slimy conidial masses. Conidia hyaline, simple, clavate, cylindrical or ellipsoidal, usually biguttulate, globose to cylindrical and mostly aggregated in slimy heads at the apex of each phialide.

2. ***Alternaria*** Nees, System der Pilze und Schwämme: 72 (1816) [1816–1817] (= *Ulocladium* Preuss)

Diagnostic characters Colonies effuse, gray, blackish brown to olivaceous-black or grayish to thick brown, fast growing, and are suede-like to floccose. Mycelium immersed or partly superficial, hyphae, stroma rarely formed. Conidiophores brown, macronematous, simple or irregularly or loosely branched, solitary or fasciculate, erect or somewhat decumbent, usually colored, straight or somewhat irregularly bent, with a single terminal scar, or sometimes with lateral scars also, often geniculate with scars at the geniculations, continuing growth laterally or subterminally to the scar, but never directly through the scar, not zigzag. Microscopically, branched acropetal chains (blastocatenate) of multicellular conidia (dictyoconidia) are produced sympodially from simple, sometimes branched, short or elongate conidiophores. Conidia pale brown, typically with both cross and longitudinal septa, obclavate to obpyriform, sometimes ovoid or ellipsoidal, often with a short conical or cylindrical beak, frequently borne acropetally in long chains, less often borne singly and having an apical simple or branched appendage, smooth-walled or verrucose.

3. ***Arthrinium*** Kunze, Mykologische Hefte (Leipzig) 1: 9 (1817)

Diagnostic characters Colonies compact or disperse, black or dark brown, usually velvety, round, oval or irregular in shape. Mycelia partially superficial and submerged. Superficial mycelium hyaline when young and finally brown and septate, composed of closely interwoven and anastomosing hyphae. Conidiophores basauxic, macronematous and mononematous, simple, slender, divided into a number of hyaline cells by dark brown, conspicuous transverse septa, swollen at each septum, with a swollen basal cell. Conidia or blastoconidia brown, simple, solitary, lateral and sometimes terminal, fusoid, oblong, curved to cuspidate. Colonies of different species may produce reddish-yellow pigments in culture media depending on the species.

4. ***Aspergillus*** P. Micheli ex Haller, Historia stirpium indigenarum Helvetiae inchoata 3: 113 (1768)

Diagnostic characters Colonies fast growing, white, blue, green, yellowish green, brown, fawn, ochraceous to black, zonate or azonate, velvety, floccose, lanose or funiculose. Mycelium mostly submerged, constituting of septate, branched, colorless or bright colored hyphae, or in some cases slowly becoming brown in localized submerged areas, or forming brown crusts or sclerotia. Conidiophores erect or suberect, hyaline to subhyaline, septate or non-septate, smooth or rough, becoming pale to brown with age in some species, usually becoming broader above and terminating in a globose, subglobose, hemispherical, spathulate or clavate vesicle bearing numerous phialides borne directly on them or borne on metulae which are produced on the vesicle; metulae or phialides when directly borne on the vesicles either parallel and clustered in terminal groups or radiating from the entire surface. Conidia phialospores cut of successively from the tips of the phialides by septa and forming unbranched chains arranged into radiate (globose) heads or aggregated to form columnar masses; conidia simple, hyaline to definitely yellowish green, pyriform to almost globose, catenulate, produced basipetally. *Eurotium* Link is the teleomorphic stage which was reported by Chen et al. (2014) as endolichenic fungus from *Cladina grisea*.

5. ***Aureobasidium*** Viala and G. Boyer, Revue Génerale de Botanique 3: 371 (1891)

Diagnostic characters Colonies light brown, yellow, pink or black, fast growing, spreading, smooth, usually with sparse aerial mycelium, often covered with slimy masses of conidia. Hyphae hyaline, thin walled when young, older hyphae thick walled with thick septa, often appearing double walled, segmenting into arthroconidia or arthroconidia-like hyphal segments, usually producing also thick walled chlamydospore-like cells or *Papulospora*-like aggregations of cells, all cells capable of producing conidia on quite short or sometimes somewhat elongate papillae produced from the inner wall of cells at regular intervals. Conidia hyaline, simple, blastic, produced simultaneously in dense groups, smooth. Secondary conidia common; endoconidia often present. Occasionally dark, simple to 1 septate arthroconidia are formed.

6. ***Biatriospora*** K.D. Hyde and Borse, Mycotaxon 26: 263 (1986)

Diagnostic characters Colonies medium gray to dark gray to dark grayish yellowish brown to dark greenish gray, plane to floccose to funiculose with heaped and lanose central part which may be hyaline to blackish green, effuse with slightly ruffled margin, either producing soluble reddish or dark brown pigment or not, reverse greenish black to dark grayish reddish brown to brownish black. Mycelium sterile, smooth to granulose, hyaline to light gray to brownish gray to gray colored, sometimes producing mycelia tufts, sparsely branched, often fragmenting.

7. ***Bipolaris*** Shoemaker, Canadian Journal of Botany 37(5): 882 (1959)

Diagnostic characters Colonies effuse, blackish-brown, velvety, mycelium immersed in substrata, reverse black. Conidiophores reddish brown to brown, septate, single, indeterminate, straight or flexuous, growing sympodially, geniculate, smooth, cylindrical, perforate at apex. Conidia phaeophragmospores, pale to dark brown, 3–4 pseudoseptate, solitary, acropleurogenous, smooth or verruculose, fusoid, straight or slightly curved, clavate to fusiform to ellipsoidal, rounded at both ends.

8. ***Bispora*** Corda, Icones fungorum hucusque cognitorum 1: 9 (1837)

Diagnostic characters Colonies punctiform or effuse, usually fuscous or black. Mycelium immersed or sometimes partly superficial, creeping, septate, yellowish to dark brown. Conidiophores hyaline to pale brown or brown, semi-macronematous, mononematous, usually inconspicuous and short on natural substrates, often longer in culture, straight or flexuous, simple or sparingly branched, smooth. Conidia dark brown or black, 1 to less often 2 septate, with a very dark brown to black band at the septum, catenulate, acrogenous, doliiform or cylindrical, rounded at the ends, produced in long unbranched acropetal chains.

9. ***Botrytis*** P. Micheli ex Haller, Historia stirpium indigenarum Helvetiae inchoata: 111 (1768)

Diagnostic characters Colonies effuse, often gray, powdery; under the low-power binocular microscope stout brown conidiophores are seen supporting glistening heads of pale conidia. Mycelium immersed or superficial, hyphae hyaline, branched, septate, creeping. Sclerotia frequently formed. Conidiophores brown, septate, macronematous, mononematous, straight or flexuous, smooth, simple or frequently dichotomously or trichotomously branched, with branches mostly restricted to the apical region forming a stipe and a rather open head (tree-like appearance); the apical cells of the branches are enlarged or rounded, bearing clusters of conidia on short sterigmata. Conidia hyaline or ash-colored, gray in mass, simple, smooth, solitary, acropleurogenous, ellipsoid to obovoid to spherical or subspherical, apiculate at the base.

10. ***Broomella*** Sacc., Sylloge Fungorum (Abellini) 2: 557 (1883)

Diagnostic characters Acervuli black, solitary to aggregated, pycnidium-like, globose to conical. Conidiophores erect, hyaline, with 1–2 septa, branched at the base, short. Conidia dark brown, triseptate, fusoid to subclavate, straight or curved, widest above the middle septum; central cells dark brown, warty, thick walled; end cells hyaline, thin walled; apical cells bearing 2–4 (usually 3 or 2), flexuous, hyaline unbranched appendages; basal cell conical, truncate at the end, with a short, central endogenous appendage.

11. ***Chaetomella*** Fuckel, Jahrbücher des Nassauischen Vereins für Naturkunde 23–24: 401 (1870)

Diagnostic characters Mycelium superficial or immersed, hyaline to pale brown, branched, septate. Conidiomata pycnidial. Pycnidia globose, superficial, dark brown to carbonaceous, pale when young, darker with age, unilocular, thick walled. Conidiophores hyaline, simple to branched, filiform, septate, smooth producing conidia acropleurogenously. Conidia dark to subhyaline, simple, fusoid to somewhat curved.

12. ***Chaetomium*** Kunze, Mykologische Hefte (Leipzig) 1: 15 (1817)

Diagnostic characters Colonies rapidly growing, cottony, initially white in color, mature colonies gray to olive in color, reverse tan to red or brown to black. Hyphae upright, simple to septate. Conidiophores short, irregularly branched, hyaline, bearing a loose cluster of conidia. Conidia simple, hyaline, borne singly, globose. It is teleomorphic stage of *Botryotrichum* Sacc. & Marchal.

13. ***Chalara*** (Corda) Rabenh., Deutschlands Kryptogamenflora (Leipzig) 1: 38 (1844)

Diagnostic characters Colonies effuse, gray, olive, brown or black, hairy or velvety. Mycelium immersed or superficial, dark. Conidiophores macronematous, mononematous, straight or slightly flexuous, unbranched, brown, smooth. Conidia hyaline to rarely brown, simple to 3 septate, cylindrical or oblong with truncate ends, catenate, endogenous, smooth or with the ends verruculose, often hanging together in chains. It is the anamorphic stage of some *Cylindrocephalum* Bonord. species.

14. ***Chrysosporium*** Corda, Deutschlands Flora, Abt. III. Die Pilze Deutschlands 3–13: 85 (1833)

Diagnostic characters Colonies flat, white to cream-colored with a very granular surface, reverse pale or pale brownish-yellow with age. Conidiophores simple, hyaline, bearing conidia at the tips of short or long lateral branches or sessile along the hyphae (intercalary). Conidia numerous, hyaline, simple, clavate to pyriform, smooth, slightly thick-walled, and have broad truncate bases and pronounced basal scars.

15. ***Cladosporium*** Link, Magazin der Gesellschaft Naturforschenden Freunde Berlin 8: 37 (1816)

Diagnostic characters Colonies effuse, olivaceous, buff, dark brown. Mycelium well developed, submerged, velvety, floccose, hairy, composed of hyaline, thin walled, narrow, cylindrical, long celled hyphae at first; later becoming brownish, thick walled, inflated, short celled. Conidiophores septate, olivaceous to dark brown, macronematous, arising as scattered, lateral or terminal outgrowths from the hyphe or as fascicles from dark brown, pseudoparenchymatous stromata, straight or flexuous, usually unbranched or branched variously near the apex or middle portion, clustered or single, smooth or verrucose. Conidia subhyaline to dark olivaceous brown or brown, simple to 2 or more septate, catenate, single or in simple or branched chains, forming tree-like heads, cylindrical, doliiform, ellipsoidal, fusiform, oval with distinctly protuberant scar at each end or just at the base, smooth, verrucose or echinulate.

16. ***Colletotrichum*** Corda, Deutschl. Flora, III (Pilze): 41, tab. (1831)

Diagnostic characters Colonies usually darkly pigmented with white aerial mycelium, consisting of numerous black sclerotia and light brown-colored conidial masses, reverse side dark brown. Sclerotia abundant, confluent, setose, spherical. Conidiophores simple, elongate. Conidia hyaline, simple, ovoid or oblong, attenuated at the ends. It's teleomorphic stage is *Glomerella* Spauld. & H. Schrenk which is being isolated by Suryanarayanan et al. (2005) as endolichenic fungus from *Physcia aipolia*.

17. ***Coniochaeta*** (Sacc.) Cooke, Grevillea 16 (77): 16 (1887)

Diagnostic characters Colonies flat, smooth, moist, pink to orange, with regular and sharp margin; reverse side pink. Hyphae narrow, hyaline, producing conidia laterally from small collarettes directly on the hyphae, or from lateral cells which are sometimes arranged in dense groups; lateral cells flask-shaped or nearly cylindrical. Collarettes unpigmented. Conidia hyaline, smooth and thin-walled, broadly ellipsoidal to cylindrical or allantoid, produced in slimy heads.

18. ***Corynespora*** Güssow, Zeitschrift für Pflanzenkrankheiten und Pflanzenschutz 16: 10 (1906)

Diagnostic characters Colonies effuse, gray, olivaceous brown, brown, dark blackish brown, black, often hairy or velvety. Mycelium immersed or superficial. Conidiophores macronematous, mononematous, straight or flexuous, unbranched, brown or olivaceous brown, smooth. Conidia solitary or catenate, dry, acrogenous, simple, obclavate in most species, cylindrical in few, subhyaline, pale to dark brown or olivaceous brown or straw colored, septate or pseudoseptate, smooth to verrucose.

19. ***Curvularia*** Boedijn, Bulletin du Jardin Botanique de Buitenzorg 13 (1): 123 (1933)

Diagnostic characters Colonies are fast growing, suede-like to downy, brown to blackish brown with a black reverse. Mycelium composed of branched, septate, subhyaline to brown hyphae. Conidiophores brown, erect or decumbent, straight, bent or flexuous, usually simple, septate, sometimes branched, geniculate towards the apex, bearing conidia successively on new growing tips. Conidia olivaceous to dark brown, with three or more transverse septa (phragmoconidia) and are formed singly and acrogenously at the tip of elongating geniculate conidiophore through a pore (poroconidia) in a sympodial manner. Conidia are cylindrical or slightly curved, with one of the central cells being larger and dark colored, while end cells paler.

20. ***Cylindrocarpon*** Wollenw., Phytopathology 1: 225 (1913)

Diagnostic characters Colonies fast growing, hyaline or bright-colored, suede-like or woolly. Mycelium white, yellowish, citrine, rose-carmine-violet. Sporodochia may occasionally be present, globose, smooth or rough, hyaline, yellow, golden, red-violet, ochraceous, brown, rarely green or black. Conidiophores simple, or irregularly or alternately or verticellately branched, free, floccose, caespitose or sporodochial, terminating in phialides. Phialides are cylindrical to subulate, with small collarettes producing hyaline, smooth-walled conidia, which are arranged ins-limy masses. Conidia of two types a) macroconidia – simple to several septate, hyaline singly but in mass white, cream, yellow, pink, rose, abstricted singly from the tips of the phialides, cylindrical, cylindric-fusiform, claviform, straight or curved, sometimes obtuse, obtuse-conic, more or less convex, sometimes apiculate at the base, sometimes pedicellate, scattered or in false heads or in tuberculate sporodochia, b) microconidia – simple, ovate, fusiform or pear shaped. Chlamydospores may be present or absent, simple, in chains or clusters, usually intercalary, rarely terminal hyaline to brown, spherical. It is the anamorphic stage of *Neonectria* Wollenw.

21. ***Cylindrosporium*** Grev., Scottish Cryptogamic Flora (Edinburgh) 1: pl. 27 (1822)

Diagnostic characters Mycelium immersed, branched, septate, hyaline. Conidiomata acervulus, white, slimy, separate or confluent, formed of pale brown to hyaline, thin-walled textura angularis. Conidiophores short, hyaline, parallel, branched only at the base, 1–2 septate, smooth. Conidia hyaline, simple to septate, filiform, straight or slightly curved, thin-walled, smooth.

22. ***Drechslera*** S. Ito, Proceedings of the Imperial Academy Japan 6 (8): 355 (1930)

Diagnostic characters Colonies fast growing, effuse, gray, brown or blackish brown with a black reverse, often hairy, sometimes velvety. Mycelium mostly immersed. Stroma present in some species. Sclerotia or protothecia often formed in culture. Conidiophores brown, septate, often geniculate, macronematous, mononematous, sometimes caespitose, straight or flexuous, unbranched or in a few species loosely branched, smooth in most species. Conidia phaeophragmospores, straw-colored or pale to dark brown or olivaceous brown, simple, solitary, in certain species also sometimes catenate or forming secondary conidiophores which bear conidia, acropleurogenous, straight or curved, clavate, ellipsoidal, fusiform or obclavate, rounded at the ends, sometimes with cells unequally colored, the end cells then being paler than intermediate ones, mostly smooth, rarely verruculose, pseudoseptate.

23. ***Epicoccum*** Link, Magazin der Gesellschaft Naturforschenden Freunde, Berlin 8: 32 (1815)

Diagnostic characters Colonies are fast growing, suede-like to downy, with a strong yellow to orange-brown diffusible pigment. Sporodochia dark, more or less cushion shaped, variable in size. Conidiophores compact or loose, hyaline to dark, rather short. Conidia darkly pigmented, simple to multicellular (dictyoconidia), globose to pyriform, mostly with a funnel-shaped base and broad attachment scar, often seceding with a protuberant basal cell.

24. ***Fusarium*** Link, Magazin der Gesellschaft Naturforschenden Freunde, Berlin 3 (1): 10 (1809)

Diagnostic characters Colonies usually fast growing, cottony, with some tinge of pink, purple or yellow color. Aerial mycelium felty or diffuse. Conidiophores hyaline, variable, septate, slender, simple or branched once or repeatedly, terminating in phialides which are sometimes formed verticellately. Phialides variable in size and shape, but mostly subulate. Conidial masses typically formed in sporodochia or in pionnotes or sometimes scattered in the mycelium. Conidia hyaline, of two kinds, often held in a mass of gelatinous material, a) macroconidia – hyaline, 3- to many septate, slightly curved or bent at the pointed ends, fusiform to falcate, dorsiventral, straight or curved, frequently naviculate (boat shaped) with foot cell, produced in mucus or slimy masses, b) microconidia – simple, ovoid or oblong, borne singly at the tips of phialides or else abstracted in succession at the tips of the phialides to form simple chains. Some conidia intermediate, 1- or 2-septate, oblong or slightly curved.

25. ***Geniculosporium*** Chesters and Greenh., Transactions of the British Mycological Society 47 (3): 400 (1964)

Diagnostic characters Colonies grayish, occasionally somewhat brownish, generally with a low and floccose surface, and may be irregularly zonate. Mycelium immersed or superficial. Conidiophores smooth-walled, rather brownish except in the apical fertile regions where they are usually colorless, erect and branched, each branch arising just beneath a septum. The parent conidiophores may be slightly displaced to one side; then the branching may superficially appear to be dichotomous and often two or more branches arise beneath the same septum; thus, there is no morphologically well-defined main axis. Conidia simple, hyaline, smooth, sub-globose to ovoid-ellipsoidal, aerogenous and produced in continuous succession at the apex. *Nemania* Gray is the teleomorphic stage of this genus.

26. ***Geomyces*** Traaen, Nytt Magazin for Naturvidenskapene 52: 28 (1914)

Diagnostic characters Colonies only slightly spreading, initially white but soon becoming pale brown, dark brown, gray, reddish or yellow, felty or powdery, often with scattered tufts of aerial hyphae. Hyphae hyaline or pale yellow, thin-walled, narrow, branching of sterile hyphae more or less orthotropic and usually not correlated with the septation of the supporting hypha, branching of fertile hyphae more frequent, at acute angles, often once or twice verticillate with 2–4 branches per whorl, the secondary or tertiary branches arising just below the septa of the primary or secondary branch respectively. Thallic conidia normally borne terminally on verticillate branches, intergrading with intercalary conidia, or borne laterally and solitary on short protrusions or short side branches, subhyaline, pale yellow or pale greenish, smooth-walled or echinulate, thin-walled to slightly thick-walled, cuneiform, obovoid, ellipsoid or clavate, simple, truncate with wide basal scars. Intercalary conidia borne on the outer branches of the verticillate hyphae, alternate, separated by short, mostly more broad than long sterile hyphal segments, in series of 2–4, subhyaline, pale greenish or pale yellow, smooth-walled or echinulate, thin-walled to slightly thick-walled, barrel-shaped, simple, broader than the supporting hypha, truncate at both ends.

27. ***Geopyxis*** (Pers.) Sacc., Sylloge Fungorum (Abellini) 8: 63 (1889)

Diagnostic characters Mycelial mats covered by persistently white and cottony mycelium, underneath consisting of a leathery to cartilaginous mat of densely interwoven hyphae which are smooth, hyaline to pale-brown hyphae with a blister-like ornamentation. Conidiophores branched, hyaline, smooth to incrustrated carrying 10–20 minutely warted, ochraceous, spherical, oval or pyriform conidia. Chlamydospores typically clavate, smooth, produced on tips of hyphal bundles adpressed to or submerged in the growth medium.

28. ***Geotrichum*** Link, Magazin der Gesellschaft Naturforschenden Freunde, Berlin 3 (1): 17 (1809)

Diagnostic characters Colonies thin, spreading, creamy white, farinose or hairy, soft and somewhat yeast-like texture. Hyphae hyaline, septate, more or less specialized into broad, radiating vegetative hyphae branching dichotomously, and narrower, lateral, sporulating hyphae which may also branch. Arthroconidia formed in chains by the disarticulation of the lateral hyphae at septa, sometimes intercalary in the broad vegetative as well, hyaline, simple, smooth, subglobose to cylindrical with truncate ends. Conidia (oidia) may also develop sympodially and chlamydospores and endoconidia may also be present. The arthroconidia, which are quite variable in size, may germinate at one end giving the appearance of a bud. However, the latter develops into a septate mycelium. It is the anamorphic stage of *Dipodascus* Lagerh.

29. ***Gilmaniella*** G.L. Barron, Mycologia 56: 514 (1964)

Diagnostic characters Colonies effuse, at first pale gray, later dark blackish-brown; mycelium superficial and immersed; hyphae smooth, hyaline when young, later becoming brown and verruculose or finely echinulate, tranverse septa often thick and very dark. Conidiophores semi-macronematous, mononematous, straight or flexuous, frequently branched, rarely unbranched, colorless, smooth. Conidiogenous cellsmonoblastic, polyblastic, integrated, terminal, intercalary or discreate, determinate, cylindrical. Conidia solitary, often in botryose clusters, dry, acropleurogenous, simple, spherical, dark brown with a small but very distinct germ pore, smooth.

30. ***Helminthosporium*** Link, Magazin der Gesellschaft Naturforschenden Freunde, Berlin 3 (1): 10 (1809)

Diagnostic characters Colonies effuse, dark, hairy. Mycelium immersed. Conidiophores macronematous, mononematous, unbranched, often caespitose, straight or flexuous, cylindrical or subulate, mid to very dark brown, smooth or occasionally verruculose, with small pores at the apex and laterally beneath the septa. Conidia solitary, rarely catenate, acropleurogenous, developing laterally often in verticils through very small pores beneath the septa whilst the tip of the conidiophores is actively growing, growth of the conidiophore ceasing with the formation of terminal conidia, simple, usually obclavate, sometimes rostrate, subhyaline to brown, smooth, pseudoseptate, frequently with a prominent, dark brown or black scar at the base.

31. ***Heteroconium*** Petr., Sydowia 3 (1–6): 264 (1949)

Diagnostic characters Colonies effuse, cottony, olivaceous to dark reddish brown, thin. Mycelium partially superficial, partially immersed. Conidiophores solitary, usually simple, mid to dark brown or reddish brown, often swollen near the base, macronematous, mononematous, unbranched, straight or curved, smooth. Conidia mid to dark brown or reddish brown, simple to 4-septate (most commonly 1 or 3), smooth, in long, simple, often spirally twisted chains, catenate, dry, acrogenous, cylindrical to fusiform, usually slightly curved with rounded ends or obclavate.

32. ***Humicola*** Traaen, Nytt Magazin for Naturvidenskapene 52: 31 (1914)

Diagnostic characters Colonies effuse, velvety, sometimes funiculose, at first white, later pale gray, grayish-brown or blackish-brown. Mycelium superficial and immersed, hypha hylanie, yellow brown in mass, septate, branched. Conidiophores erect, straight, short, sometimes septate, simple or rarely branched, colorless to pale golden brown, bearing a single conidia at the apex. Conidia pale to golden brown, simple, solitary, subglobose to globose, usually smooth. Some species also produces phialides directly from aerial mycelium. Phialides hyaline, subulate and sinuous, simple, tapering upward, delimited by a septum from the parent hypha, producing small, ovoid phialospores in chains. Phialospores small, hyaline, simple, obovoid, abstracted from the tip of the phialide in succession, aggregated in balls or forming simple chains.

33. ***Karsteniomyces*** D. Hawksw., Transactions of the British Mycological Society 74 (2): 371 (1980)

Diagnostic characters Conidiomata pycnidial, arising singly, scattered or loosely aggregated, subglobose, superficial, nectrioid, translucent pale orange to deep red, ostiole irregular and forming schizogenously. Conidiophores cylindrical, sympodially branched. Conidiogenous cells holoblastic, arising acrogenously or pleurogenously, cylindrical, hyaline. Conidia probably dry, elongate-ellipsoid, the apex rounded and the base slightly to clearly truncate, hyaline, 1 septate, smooth-walled, guttulate.

34. ***Lasiodiplodia*** Ellis and Everh., Botanical Gazette Crawfordsville 21: 92 (1896)

Diagnostic characters Colonies grayish sepia to mouse gray to black, fluffy with abundant aerial mycelium; reverse fuscous black to black. Conidiophores are hyaline, simple, sometimes septate, rarely branched cylindrical, arising from the inner layers of cells lining the pycnidial cavity. Conidia hyaline to cinnamon to fawn, simple to 1-septate, granulose, subovoid to ellipsoid-oblong, thick-walled, base truncate.

35. ***Massarina*** Sacc., Sylloge Fungorum (Abellini) 2: 153 (1883)

Diagnostic characters Colonies slow to fast growing, forming a whitish to yellowish to reddish purple to copious blackish gray to blackish brown aerial mycelium; conidia hyaline, simple, ellipsoidal. Anamorphs formed in culture or found in nature in close association, belongs to *Acrocalymma* Alcorn and J.A.G. Irwin, *Chaetophoma* Cooke, *Diplodia* Fr., *Periconia* Tode, *Phoma* Sacc., *Stagonospora* (Sacc.) Sacc., *Tetraploa* Berk. & Broome. Generally it is a pyrenocarpous fungus characterized by pseudoseptate parahyses, ascospores fusifrom to long-ellipsoid, hyaline, asymmetrically or symmetrically 1–3(−7) septate, often surrounded by a thin or thick gelatinous sheath and/or with polar gelatinous appendages and bitunicate asci.

36. ***Monilia*** Bonord., Handbuch der allgemeinen Mykologie (Stuttgart): 76 (1851)

Diagnostic characters Colonies compact, whitish, later flesh colored to ochraceous. Mycelium composed of creeping, septate, hyaline, branched hyphae. Conidiophores mostly fasciculate, septate, hyaline, arising laterally or terminally on hyphae, ascending or erect, dichotomously or racemosely or irregularly branched, branching limited or abundant. Conidia hyaline or pink, gray or tan in mass, simple, ovate to elongate or somewhat subglobose, seldom globose, catenulate, formed in long, branched chains in bead like fashion, often united by isthmus-like connecting joints, produced in basifugal succession.

37. ***Mucor*** Fresen., Beiträge zur Mykologie 1: 7 (1850)

Diagnostic characters Colonies fast growing, white to yellow, becoming dark gray with the development of sporangia. Mycelium present without rhizoids and stolon. Sporangiophores colorless or colored, forming a thick turf, springing from the mycelium, either unbranched with terminal sporangia or racemosely branched with sporangia on all the branched ends, monopodial. Sporangia erect on sympodial sporangiophores, non apophysate. Columellae large, colorless or grayish or brownish. Zygospores produced between compatible types in the aerial myceliun dark brown, with characteristic more or less stellate warts, mostly heterothallic.

38. ***Myxotrichum*** Kunze, Mykologische Hefte (Leipzig) 2: 108 (1823)

Diagnostic characters Being a member of Gymnoascaceous fungi, it is characterized by dark colored, bramble-like ascocarps; centrum at first white then yellowish at marurity; ascospores hyaline or light colored, fusiform, lenticular, elliptical, cymbiform or navicular, occasionally with a hyaline rim around the longitudinal axis. Anamorphic stages represented by intercalary, more or less oblong or cylindrical arthro-aleuriospores or by spherical or clavate, terminal aleuriospores. Chlamydospores occasionally present.

39. ***Nectria*** (Fr.) Fr., Summa vegetabilium Scandinaviae (Stockholm) 2: 387 (1849)

Diagnostic characters Colonies whitish yellow to whitish luteous to yellowish brown, with a fruity odor. Lateral phialidic pegs are produced after 2–3 days with abundant conidia. Phialidic pegs are ellipsoidal and slightly tapering toward the tip or flask shaped. Conidiophores may be aerial or form sporodochia. Aerial conidiophores are unbranched, sometimes verticillate, 1–3- branched, becoming loosely to moderately branched. Sporodochia are generally solitary, occasionally caespitose, convex to concave. The sporodochial conidiophores are 2–3 branched, becoming densely branched with terminal whorls of 2–4. Conidia acropleurogenous, produced on monophialides on aerial, submerged or repent hyphae, smooth, hyaline to whitish yellow to saffron when fresh and sienna to red to scarlet when dry, ellipsoidal, oblong to cylindrical, subglobose to obovate, sometimes allantoids, straight or slightly curved. Chlamydospores are globose, subglobose or broadly ellipsoid, simple to 1 septate.

40. ***Neurospora*** Shear & B.O. Dodge, Journal of Agricultural Research, Washington 34 (11): 1025 (1927)

Diagnostic characters Colonies white, with scanty aerial hyphae especially at the periphery of the plate; margins smooth or partly irregular; reverse dark. Hyphae hyaline, septate, thick-walled, branched. Perithecia scattered to aggregated, pale brown to dark brown to black colored,smooth or downy with loose hyphae. Asci unitunicate, hyaline, cylindrical, thin-walled having ring-like thickening at the tip, short stalked, often 8-spored. Ascospores uniseriate or somewhat overlapping, initially hyaline, becoming yellowish brown to black with maturity, simple, ellipsoidal or elongate, ascospore wall surface with global ornate pits and abnormal hyaline sheath.

41. ***Nigrospora*** Zimm., Centralblatt für Bakteriologie und Parasitenkunde 8: 220 (1902)

Diagnostic characters Colonies fast growing, gray, tomentous. Mycelium composed of creeping, septate, branched hyphae which are at first hyaline and turn brown later. Conidiophores brownish, short, ampulliform, cells somewhat inflated, mostly simple, rarely slightly branched, inflated near the tip, bearing a single conidium at the tip. Conidia shiny, initially pale yellow to orange yellow and at maturity olive brown to black, simple, globose, sometimes with a slight apical bulge, situated on a flattened, hyaline vesicle at the end of the conidiophores, typically quite round when viewed from the end and elliptical when seen from the side.

42. ***Nodulisporium*** Preuss, Klotzschii herbarium vivum mycologicum sistens fungorum per totam Germaniam crescentium collectionem perfectam. Editio prima. Centuria XIII: no. 1272 (1849)

Diagnostic characters Colonies plane and velutinous, with irregular margins and mycelium subsurface, whitish felt-like mycelia on the periphery while distinctively grayish/brown to pale brown in the centre. Reverse deeply colored dark brown, near dark mouse gray. Mycelium composed of septate and slightly roughened hyphae. Conidiophores erect to suberect, smooth or slightly roughened like the mycelium and often studded with dark granules, mononematous (coindiophores solitary), irregularly branched, flexuous (wavy, not rigid), septate, and becoming mid-brown in the apical part with age. Conidia arose singly and successively on denticles at the tips of conidiogenous cells; the first conidium was formed apically. Subsequent conidia were formed sympodially in more or less basipetal succession, forming heads. Conidia pale brown and smooth, dry, simple, ellipsoid to clavate ellipsoid, with a flattened base indicating the former point of attachment to conidiogenous cell (Umabala et al. 2001). The conidium production on areas of the colony surface, the conidiophores, and conidiogenesis were similar to those occurring in many anamorphs of *Hypoxylon* Bull.; *Daldinia* Ces. & De. Not. more specifically they were similar to those occurring in the genus *Nodulisporium*. The anamorphic stage *Hypoxylon* was isolated as endolichenic fungus by Li et al. (2007) from *Punctelia borreri* and by Vinayaka et al. (2016) from *Cladonia fruticulosa* and *Parmotrema reticulatum*.

43. ***Paecilomyces*** Bainier, Bulletin de la Société Mycologique de France 23(1): 27 (1907)

Diagnostic characters Colonies fast growing, powdery or suede-like, white or pale, gold, green-gold, yellow-brown, lilac or tan, but never green or blue-green as in *Penicillium*. Conidiophores mostly arising from aerial hyphe, sometimes absent. Phialides short-tubular, in a loose verticillate group on the conidiophores or arising directly from the mycelium, more divergent than in *Penicillium*; basal portion of phialide nearly cylindrical, tapering gradually or abruptly to a long slender tube. Conidia produced in basipetal succession from the phialides and held together in long, dry chains or in irregular masses under moist conditions, simple, hyaline to dark, smooth or rough, ovoid to fusoid.

44. ***Papulospora*** Preuss, Linnaea 24: 112 (1851)

Diagnostic characters
Colonies variously colored on different media: colorless to white (on CMA), green on MEA and PDA, and gray-brown on SAB and mycosel agar. Hyphal development is luxuriant and superficial to immersed. Bulbils start forming at the perimeter of

each colony, both on the surface and subsurface, and later on spreads to the entire colony. Mature bulbils are elongate to elliptical to oval, with irregularly lobed surface. When crushed, the mature structures appear to be amorphous, disintegrating into individual fragments of swollen cells with no fertile regions noted. Individual bulbils are produced from intertwined, anastomosed hyphae which initiate as hooks at the end of short lateral branches. The hyphae then coiled tightly, crozier fashion by [1] coiling on themselves, [2] coiling on the parent integrated hyphae, or less often, [3] on another lateral or terminal hypha. Meanwhile, other branches wrap around the basal hyphae knotting to form a bulbil initial. The inner strands of hyphae divide into structures most resembling arthroconidia or chlamydospores, and one to eight or ten of these terminal cells enlarge more rapidly and become the bulbil centers around which the remaining cells from a ring, attaching firmly and completely covering the cortex. Germination has not been observed. It is an orthographic variant of *Papulaspora Preuss*.

45. ***Penicillium*** Link, Magazin der Gesellschaft Naturforschenden Freunde, Berlin 3 (1): 16 (1809)

Diagnostic characters Colonies dense, floccose, variable in color, white, green, yellow-green, blue-green, gray, or less often colorless or avellaneous to yellowish, reddish, purplish, or other tints, frequently becoming brownish in mass when mature. Soluble pigment pale grayish red; exudates, clear to pale yellow; reverse side yellow, to light orange becoming dark orange to reddish orange with age. Hyphae creeping, branched, septate, most commonly smooth or occasionally rough, hyaline to subhyaline, pale yellow brown in mass. Conidiophores usually conspicuous, hyaline, rough or smooth, septate, arising as branches from the vegetative hyphae, commonly at right angles to the parent hypha, sometimes branched at or near the apex, branches divergent or adpressed to the main conidiophores axis. Conidial structures brush- or broom-like and forming a peniculus ranging from a single terminal vertical of conidia-bearing phialides or a terminal vertical of equal branches or metulae bearing verticils of phialides, to complex branching systems terminating in verticils of specialized cells or metulae which bear clusters of phialides. Phialides ampulliform or acerose, hyaline, borne in groups directly at the apex of conidiophores, or branches of the conidiophores. Conidia hyaline or sometimes colored individually or in mass, but not dematiaceous, simple, abstricted from the tips of the phialides successively and forming unbranched, basipetal chains, cylindrical to elliptical, oval or globose.

46. ***Periconia*** Tode, Fungi Mecklenburgenses Selecti (Lüneburg) 2: 2 (1791)

Diagnostic characters Colonies effuse, dark olivaceous brown, purplish brown or black, velvety. Mycelium partly superficial and partly immersed, composed of hyaline to pale brown, smooth-walled, branched, septate hyphae. Conidiophores macronematous, mononematous, simple, long, narrow, straight or flexuous, mid to dark brown or reddish brown, closely verruculose or echinulate, bearing rather short, smooth, fertile, widely spaced, often unciform lateral branches. Conidia blastosporous, simple, dark brown, developed acropetally, spherical to globose, verruculose or minutely echinulate.

47. ***Pestalotiopsis*** Steyaert, Bulletin du Jardin Botanique de l'État à Bruxelles 19 (3): 300 (1949)

Diagnostic characters Colonies compact or effuse, buff, grayish brown, blackish brown or black; mycelium immersed, branched, septate, hyaline to pale brown. Conidiomata acervulii, separate or confluent, formed of brown, thin walled textura angularis. Conidiophores hyaline, branched and septate at the base and above, cylindrical or lageniform. Conidiogenous cells holoblastic, annellidic, indeterminate, integrated, cylindrical, hyaline, smooth, with several percurrent proliferations. Conidia fusiform, straight or slightly curved, 4-septate; basal cell hyaline, truncate, with an endogenous, cellular, simple or rarely branched appendage; apical cell conic, hyaline, with two or more apical, simple or branched, spathulate or espathulate appendages; median cells brown, sometimes versicolored, thick walled, smooth or verruculose.

48. ***Peziza*** Fr., Systema Mycologicum 2: 40 (1822)

Diagnostic characters The conidiophores usually 2 to 14 septate, and end in an *Oedocephalum* head which varies greatly in shape and size, narrow clavate, oval to spherical. Conidia hyaline, simple, with a slightly roughened or warted surface, elliptical. The point where the conidium was originally attached usually shows as a little papilla on the conidiophore head, and as a little knob on the lower end of the conidium.

49. ***Phaeosphaeria*** I. Miyake, Botanical Magazine, Tokyo 23: 93 (1909)

Diagnostic characters Colonies erumpent, spreading, flat surface; margins uneven, feathery; aerial mycelium absent to sparse; dirty white to pale luteous to luteous to olivaceous gray with reverse luteous to olivaceous gray to saffron. Mycelium immersed, composed of branched, septate, hyaline hypahe. Aerial hyphae formed at the centre of the colonies. Ascomata immersed to partially immersed, pyriform, scattered, papillate, ostiolate, dark brown. Peridium pseudoparenchymatic. Asci clavate, slightly curved toward the base, fissitunicate, 8-spored. Ascospores biseraiately arranged, hyaline, broadly fusiform, straight or slightly curved, 3-septate, surrounded by gelatinous sheath which do not stain with Indian ink.

50. ***Phialophora*** Medlar, Mycologia 7 (4): 202 (1915)

Diagnostic characters Colonies are usually slow growing, gray to olivaceous-black, often becoming brown with age. Conidiophores dark, short or absent; phialides variable in shape but usually broader near the middle and tapering toward both ends, and producing conidia endogenously. Conidia subhyaline to olivaceous brown, smooth-walled, ovoid to cylindrical or allantoid, simple, usually aggregate in slimy heads at the apices of the phialides, which may be solitary, or in a brush-like arrangement.

51. ***Phoma*** Sacc., Michelia 2(6):4 (1880)

Diagnostic characters Colonies spreading, grayish-brown, powdery or suede-like and produce large, globose, membranous to leathery, brown to dark brown, ostiolate pycnidia. Mycelium branched, septate, hyaline or pale brown. Conidiophores short or obsolete. Conidia arising singly, not catenate, produced in abundance within the pycnidia on narrow thread-like phialides, which are hardly differentiated from the inner pycnidial wall cells. Conidia simple when mature (exceptionally becoming 1-septate), hyaline, ellipsoid, subcylindrical, fusiform, pyriform or globose, rounded at the apex and base or with the base slightly truncated, often guttulate.

52. ***Phomopsis*** Sacc. & Roum., Revue Mycologique Toulouse 6: 32 (1884)

Diagnostic characters Colony fluffy to slightly fluffy, compact, thick to thin, greenish to whitish to grayish to brownish, zonate to azonate. Mycelium hyaline to pale brown. Pycnidia dark, ostiolate, immersed, erumpent, nearly globose. Conidiophores branched and septate at the base and above, occasionally short and only 1–2 septate, more frequently multiseptate and filiform, hyaline. Conidia hyaline, simple, of 2 types: α conidia – ovoid to cylindrical to fusiform, straight, usually biguttulate (one guttule at each end) but sometimes with more guttules, aseptate, and β conidia – filiform, curved or bent stylospores, eguttulate, aseptate.

53. ***Phyllosticta*** Pers., Traité sur les Champignons Comestibles (Paris): 55, 147 (1818)

Diagnostic characters Conidiomata variable from pycnidial to pycnidioid, immersed, subepidermal or subperidermal in origin, unilocular or plurilocular, glabrous, ostiolate, dark brown to black; ostiole circular or oval; wall variable in thickness, of an outer textura angularis with thick-walled, dark brown cells and an inner pale brown to colorless, thin-walled textura prismatica. Conidiophores lining the cavity of the conidioma, reduced to conidiogenous cells, invested in mucus. Conidiogenous cells discrete, often of two kinds; a) those producing macroconidia and b) those producing microconidia (occasionally microconidia may be produced in separate conidiomata). Macroconidiogenous cells ampulliform, lageniform, subcylindrical, colorless, smooth. Microconidiogenous cells ampulliform to lageniform or conical, colorless, smooth. Macroconidia subglobose, ovoid or obovoid, ellipsoidal or pomiform, subcylindrical, broadly rounded at the apex, often tapering strongly toward the base, unicellular, colorless, smooth-walled, guttulate, often enclosed in a mucilaginous sheath and bearing an unbranched, attenuated, mucoid, apical appendage of type-G. Microconidia cylindrical or dumb-bell shaped with rounded or blunt ends, unicellular, colorless.

54. ***Preussia*** Fuckel, Fungi Rhenani Exsiccati Cent. XVI-XVIII 17–18: 175 (1866)

Diagnostic characters Colonies cottony to floccose, adpressed and partially submerged, light brown to black with cream to white patches. Ascomata scattered to aggregated, developed superficially or partially immersed in culture media when young. Pseudothecia globose to subglobose to pyriform, smooth, light brown to dark brown. Asci eight spored, cylindrical to clavate, broadly rounded above and gradually to abruptly tapering in to a robust stipe. Pseudoparaphyses filiform, septate and bifurcate. Asci 8 spored, cylindrical-clavate, stipitate, nonamyloid, broadly rounded above, gradually to abruptly tapering into a short and robust stipe. Ascospores 3 spetate, cells easily separable at the central septum, uniseriate or biseriate, cylindrical, hyaline to olivaceous when young and finally becoming olivaceous brown to dark brown when mature; transversely septate, constrictions at septa broad and shallow, middle cells of equal length and broader than terminal cells, provided with rounded apices; germ slit diagonal, oblique or parallel and straight to sinuous; gelatinous sheath hyaline and narrow.

55. ***Rhizoctonia*** DC., Flore française 6: 110 (1815)

Diagnostic characters Mycelium highly variable in appearance, composed at first of hyaline hyphae, later becoming yellowish to deep brown to violet; hyphae branched, with long cells, with nearly right-angled side branches constricted basally, septa between main hyphae and side branches set off from main hyphae. Monilioid cells composed of catenulate cells acropetally developed. Asexual fruiting bodies and spores absent. Sclerotia brown to dark brown to reddish black, variable in form, frequently small and loosely formed, formed among and connected by mycelial threads.

56. ***Rhizopus*** Ehrenb., Nova Acta Academiae Caesareae Leopoldino-Carolinae Germanicae Naturae Curiosorum 10: 198 (1820)

Diagnostic characters It is characterized by the presence of stolons and pigmented rhizoids, the formation of sporangiophores, singly or in groups from nodes directly above the rhizoids, and apophysate, columellate, multi-spored, generally globose sporangia. After spore release the apophyses and columella often collapse to form an umbrellalike structure. Sporangiospores are globose to ovoid, simple, hyaline to brown and striate in many species. Colonies are fast growing and cover an agar surface with a dense cottony growth that is at first white becoming gray or yellowish brown with sporulation.

57. ***Scopulariopsis*** Bainier, Bulletin de la Société Mycologique de France 23: 98 (1907)

Diagnostic characters Colonies fast growing, varying in color from white to brownish yellow to deep brown, grayish brown, or nearly black, never green, with aerial hyphae at least in part in creeping anastomosing ropes (funiculose). Conidiophores short, branched, usually produced on the funiculose hyphae, the upper portion bearing a cluster of phialides. Conidial apparatus variable, somewhat as in *Penicillium*, or consisting of varying and irregular groups of branches and annellophores, at times single, scattered on the aerial hyphae. Annellophores phialide-like in shape, tapering gradually from a basal tubular part to a conidium-bearing apex, or narrowly tubular without tapering, cutting off conidia from the apex by cross-walls. Conidia hyaline to brown in color, simple, globose to pyriform with a truncate base and rounded or pointed distal portion, catenulate, produced in basipetal succession on annellophores.

58. ***Sordaria*** Ces. & De Not., Commentario della Società Crittogamologica Italiana 1 (4): 225 (1863)

Diagnostic characters Colonies greenish brown to dark brown, spreading, producing abundant perithecia distributed all over the agar surface. Perithecia dark brown to black, pear-shaped or ovate, ostiolate apically, covered with white hairs. Peridium pale brown, pseudoparenchymatous. Asci cylindrical, truncate apically, basally narrow, 8-spored. Spores simple, dark brown, ovate, with 2–4 nodules per ascospore, apiculate or germ pored at one end, often covered with gelatinous sheath.

59. ***Spegazzinia*** Sacc., Michelia 2 (6): 37 (1880)

Diagnostic characters Colonies discrete, orbicular or effuse, dark blackish brown to black. Sporodochia orbicular, hemispherical or flattened, small, black, pulverulent. Mycelium superficial, composed of a close network of hyaline to subhyaline, septate, much branched and anastomosing hyphae. Conidiophores macronematous, usually arising endogenously from subspherical, ampulliform, cupulate or doliiform conidiophore mother cells, unbranched, straight or flexuous, narrow, subhyaline to brown, smooth or verrucose, short (microconidiophores) or long (macroconidiophores). Conidia pale to dark brown, produced singly at the tips of the conidiophores, dry, acrogenous, of two kinds: a) 3-septate, spiny spore, borne apically on a long slender conidiophores; b) 3-septate, smooth spore, borne on a short conidiophores.

60. ***Sporormiella*** Ellis & Everh., The North American Pyrenomycetes (Newfield): 136 (1892)

Diagnostic characters Colonies appressed to somewhat granular, pale white in the beginning, then becoming gray to somewhat black when old, dark gray to black where pseudothecia abundant, reverse hyaline to light gray; hyphae hyaline, rarely gray or brown, remotely to fairly closely septate, branching and frequently anastomosing. Pseudothecia subglobose to pyriform, immersed to semiimmersed when young, superficial when old, solitary to clustered in small groups, dark brown to nearly black, smooth, glabrous, ostiolate; neck small, papilliform, projecting above the substrate, smooth, glabrous; ostiole distinct to indistinct; peridium membranaceous to semicoriaceous, pseudoparenchymatous, dark brown. Asci eight-spored, cylindrical-clavate, stipitate, nonamyloid, broadly rounded above, gradually to abruptly tapering into a short stipe. Pseudoparaphyses abundant, filiform, septate. Ascospores 3-septate, biseriate, rarely uniseriate, cylindrical, straight or curved, hyaline to olivaceous brown when young, finally becoming dark brown to opaque at maturity, transversely septate, middle cells about equal in length and broader than terminal cells; terminal ones longer, slightly more cylindrical with rounded apices; gelatinous sheath hyaline, blackened with India ink.

61. ***Sporothrix*** Hektoen & C.F. Perkins, Journal of Experimental Medicine 5: 80 (1900)

Diagnostic characters Colonies are slow growing, moist and glabrous, with a wrinkled and folded surface. Some strains may produce short aerial hyphae and pigmentation may vary from white to cream to black. Conidiophores arise at right angles from thin septate hyphae and are usually solitary, erect and tapered toward the apex. Conidia are formed in clusters on tiny denticles by sympodial proliferation at the apex of the conidiophore, their arrangement often suggestive of a flower. As the culture ages, conidia are subsequently formed singly along the sides of both conidiophores and undifferentiated hyphae. Conidia are ovoid or elongated, hyaline, simple and smooth-walled. In some isolates, solitary, darkly-pigmented, thick-walled, simple, obovate to angular conidia may also be observed along the hyphae.

62. ***Talaromyces*** C.R. Benj., Mycologia 47(5): 681 (1955)

Diagnostic characters Colonies dense, floccose, variable in color, white, green, yellow-green, blue-green, gray, or less often colorless or avellaneous to yellowish, reddish, purplish, or other tints, frequently becoming brownish in mass when mature. Hyphae branched, septate, most commonly smooth or occasionally rough, hyaline to subhyaline, pale yellow brown in mass. Soluble pigment pale grayish red; exudates, clear to pale yellow; reverse side yellow, to light orange becoming dark orange to reddish orange with age. Conidiophores arising as branches from the vegetative hyphae, commonly at right angles to the parent hypha. Conidial structures brush- or broom-like and forming a peniculus ranging from a single terminal vertical of conidia-bearing phialides or a terminal vertical of equal branches or metulae bearing verticils of phialides, to complex branching systems terminating in verticils of specialized cells or metulae which bear clusters of phialides. Conidia hyaline or sometimes colored individually or in mass, but not dematiaceous, simple, abstricted from the tips of the phialides successively and forming unbranched, basipetal chains, cylindrical to elliptical, oval or globose. It is the perfect state of some species of *Penicillium* and is being reported by Jayakumar et al. (2016) as endolichenic fungi from *Lecanora* sp.

63. ***Thielavia*** Zopf, Verhandlungen des Botanischen Vereins der Provinz Brandenburg 18: 101 (1876)

Diagnostic characters Colonies spreading, white or light, rarely dark, often lanose or tufted, composed of septate, branched, usually hyaline hyphae; ascomata borne on hyphae, often covered by the aerial mycelium, spherical, non-ostiolate, glabrous, setose or tomentose, with a hyaline or brown wall composed of flattened hyphal cells (textura epidermoidea); asci parallel, in a fascicle or irregular, ellipsoidal, saccate or cylindrical, rarely clavate, with a rather persistent, thin wall, 4–8 spored; ascospores fusiform, ellipsoidal, obovate or clavate, simple, brown, with a conspicuous germ pore; conidial states mostly absent, chlamydospore-like structures often present.

64. ***Tolypocladium*** W. Gams, Persoonia 6 (2): 185 (1971)

Diagnostic characters Colonies growing slowly, initially with limited aerial mycelium, floccose, white, becoming very pale buff to cream colored. Reverse white to pale yellowish, or darker or brownish near the colony origin; some strains on Czapek's agar with dark green or bluish sectors corresponding with regions of dense conidiation. Exudate absent or produced as small, colorless droplets; most abundant on potato-dextrose-agar. Odor indistinct to pronounced, earthy, Streptomyces-like. Hyphae hyaline, smooth walled. Conidiophores hyaline, smooth walled, cylindrical, arising directly from the agar surface or as short branches from the aerial mycelium, sparingly branched; branches short cylindrical to ellipsoidal. Phialides solitary or in verticils of two to five, cylindrical or only slightly swollen basally, narrowing into a cylindrical neck. Conidia simple, hyaline, smooth walled, broadly ellipsoidal to subglobose or broadly obovoid to nuciform or acorn shaped, aggregating in small heads at the tips of the phialides. Chlamydospores not observed.

65. ***Torulomyces*** Delitsch, Ergebnisse der theoretischen und angewandten Mikrobiologie: Band I: Systematik der Schimmelpilze: 91 (1943)

Diagnostic characters Conidiophores monoenmatous, simple and erected, unbranched, more or less at right angles to the vegetative hyphae, hyaline, smooth. Phialides solitary, terminal, flask shaped, some of them have very clear long necks rarely are short and inconspicuous, smooth, hyaline. One septum is always observed at the base of phialides and another septum is usually observed at the base of conidiophores. Conidia produced enteroblastically from the phialides, forming long conidial chains with small connectives between adjacent conidia; hyaline to pale brown, and usually roughened showing large tubercules; globose to subglobose to rarely obovoid.

66. ***Trichobotrys*** Penz. & Sacc., Malpighia 15: 245 (1902)

Diagnostic characters Colonies effuse, dark olivaceous brown, velvety. Mycelium partly superficial, partly immersed, composed of branched, colorless to pale brown hyphae. Conidiophores mononematous, erect, straight or flexous, septate, dichotomously branched in the above half, fertile in the middle, dark to reddish brown, verruculose, conspicously echinulate in the below half, terminating in sterile, setiform, variously curved, pale brown to hyaline, apical branches. The conidiophore branches bear short, fertile, dark to pale brown, verruculose, widely spaced, 1–2-septate, laterals. Conidiogenous cells polyblastic, integrated, terminal to subterminal on fertile branches, elongated, denticulate in the upper half, sometimes collapsing into cupulate form. Conidia dry, catenate, usually in branched, acropetal chains, spherical, dark brown, verruculose, aseptate.

67. ***Trichoderma*** Pers., Neues Magazin für die Botanik 1: 92 (1794)

Diagnostic characters Colonies fast growing, at first white and downy, later developing yellowish-green to deep green compact tufts, often only in small areas or in concentric ring-like zones on the agar surface. Conidiophores indefinite, hyaline, consisting of an erect unbranched or branched hyphae bearing phialides laterally and terminally which are short and thick and either single or in groups. Phialides surrounded by heads, rarely by short chains, of slime spores (conidia). Conidia hyaline, simple, ovoid, borne in small terminal clusters; usually easily recognized by its rapid growth and green patches or cushions of conidia.

68. ***Trichophyton*** Malmsten, Archiv für Anatomie, Physiologie und Wissenschaftliche Medicin 1848: 14 (1848)

Diagnostic characters Colonies flat to slightly raised, mycelium cottony, granular or powdery in culture, white or with variable pigmentation, pinkish on the reverse side. Microconidia hyaline, small, simple, subspherical to ovoid, borne on sides of hyphae, single or in clusters. Macroconidia hyaline, large, septate, thin walled, clavate to fusiform, borne laterally directly on hyphae or on short pedicels.

69. ***Ulocladium*** Preuss, Linnaea 24: 111 (1851)

Diagnostic characters Colonies are rapid growing, brown to olivaceous-black or grayish and suede-like to floccose. Microscopically, numerous, usually solitary, multi-celled conidia (dictyoconidia) are formed through a pore (poroconidia) by a sympodially elongating geniculate conidiophore. Conidiophores smooth or verruculose, zigzag, pale brown to brown, bearing single conidia. Conidia golden brown to dark reddish brown and often verrucose, muriform, ellipsoid to obovoid (narrowest at the base). Now this genus has been synonymised under *Alternaria.*

70. ***Xylaria*** Hill ex Schrank, Baierische Flora (München) 1: 200 (1789)

Diagnostic characters Colony velvety to appressed, eventually forming radial strands, with white, abundant cottony mycelium reaching lid of Petri dish; azonate to faintly zonate; at first white, then covered by a layer of warty canary yellow mycelium, darkening to olive yellow, to blackish olive from the center outwards; margin white to canary yellow, minutely plumose; dark brown, long stromatic hyphae with short protuberances; reverse orange to golden. Stromata cylindrical or flabelliform, robust, salmon, sometimes darkening to light brown at base eventually branching when touching Petri dish lid, at first pale yellow to canary yellow or orange white, then darkening to black with yellowish white to cream or orange white tips, villose at base; colorless to yellowish droplets forming along length of stromata. Conidiophores forming after three to 4 weeks, in palisades at tip of stromata, hyaline, smooth, branching near base. Conidia ellipsoidal, simple, with flat basal abscission scar, hyaline, smooth.

Sterile mycelia, which are much alike in many features, encountered during the present study have been omitted out and there is usually more hope of their identification either morphologically if induced to form fructifications or by molecular means. However, the precise conditions required for fruiting are seldom known and thus the cultural environment is usually manipulated in the hope of getting fruit bodies; sometimes this is successful, but in most of the cases it leads to the formation of aggregate species.

References

Ando K, Nawawi A, Manoch L et al (1998) Three new species and a new combination in the genus *Torulomyces* from soil. Mycoscience 39:313–318
Arenal F, Platas G, Peláez F (2005) *Preussia africana* and *Preussia isabellae*, two new *Preussia* species based on morphological and molecular evidence. Fungal Divers 20:1–15
Barnett HL (1962) Illustrated genera of imperfect fungi, II edn. Burges, Minneapolis
Barnett HL, Hunter BB (1998) Illustrated genera of imperfect fungi, IV edn. APS Press, St. Paul, pp 1–218
Bissett J (1983) Notes on *Tolypocladium* and related genera. Can J Bot 61:1311–1329
Chen GD, Bao YR, Huang YF et al (2014) Three pairs of variecolortide enantiomers from *Eurotium* sp. with caspase-3 inhibitory activity. Fitoterapia 92:252–259
Dodge BO (1937) The Conidial Stage of *Peziza pustulata*. Mycologia 29(6):651–655
Ellis MB (1971) Dematiaceous hyphomycetes. CABI Publications/Commonwealth Mycological Institute, Kew/Wallingford, pp 608
Ellis MB (1976) More dematiaceous hyphomycetes. CABI Publications/Commonwealth Mycological Institute, Kew/Wallingford, pp 507
Fallah PM, Shearer CA, Crane JL (1998) Freshwater ascomycetes: *phaeosphaeria barriae* sp. nov. from Lakes in Wisconsin. Mycologia 90(6):1075–1078
Hawksworth DL (1980) Notes on some fungi occurring on *Peltigera*, with a key to accepted species. Trans Br Mycol Soc 74(2):363–386
Hawksworth DL (1981) The lichenicolous coelomycetes. Bulletin of the British Museum (Natural History) 9:1–98
Hirooka Y, Rossman AY, Samuels GJ et al (2012) A monograph of *Allantonectria*, *Nectria*, and *Pleonectria* (Nectriaceae, Hypocreales, Ascomycota) and their pycnidial, sporodochial and synnematous anamorphs. Stud Mycol 71(1):1–210
Jayakumar K, Prasad Babu G, Charan Theja P et al (2016) Biomedical potentials of *Talaromyces tratensis*–A new endolichenic fungi associated with high altitude curstose lichen *Lecanora* sp. Res J Pharm Biol Chem Sci 7(5):1412–1419
Khan RS, Cain RF (1979) The genera *Sporormiella* and *Sporormia* in east Africa. Can J Bot 57:1174–1186
Kolařík M, Spakowicz DJ, Gazis R et al (2017) *Biatriospora* (ascomycota: pleosporales) is an ecologically diverse genus including facultative marine fungi and endophytes with biotechnological potential. Plant Syst Evol 303:35–50
Li WC, Zhou J, Guo SY et al (2007) Endophytic fungi associated with lichens in Baihua mountain of Beijing, China. Fungal Divers 25:69–80
Nag Raj TR (1993) Coelomycetous anamorphs with appendage bearing conidia. Mycologue Publications, Waterloo, pp 1–1101
Mason EW (1940) On specimens, species and names. Trans Br Mycol Soc 24(2):115–125
Mason EW (1948) Some common mould. The Naturalist (Yorkshire), Jan–March, 5
Rahnama K, Habibi R (2015) First report of *Neurospora* on *Corylus avellana* in natural forest of Iran. J Yeast Fungal Res 6(4):31–36
Rodrigues KF, Leuchtmann A, Petrini O (1993) Endophytic species of *Xylaria*: cultural and isozymic studies. Sydowia 45(1):116–138
Subramanian CV (1983) Hyphomycetes: taxonomy and biology. Academic, London, pp 1–502
Suryanarayanan TS, Thirunavukkarasu N, Hariharan GN et al (2005) Occurrence of non-obligate microfungi inside lichen thalli. Sydowia 57(1):20–130
Sutton BC (1980) The coelomycetes. Fungi imperfecti with pycnidia, acervuli and stromata. Commonwealth Mycological Institute, Kew, pp 1–696
Talbot PHB (1971) Principles of fungal taxonomy. The Macmillan Press Ltd, London, pp 1–274
Umabala P, Lakshmi V, Murthy AR et al (2001) Isolation of a *Nodulisporium* species from a case of cerebral Phaeohyphomycosis. J Clin Microbiol 39(11):4213–4218

Vinayaka KS, Krishnamurthy YL, Banakar S et al (2016) Association and variation of endophytic fungi among some macrolichens in Central Western Ghats, southern India. Int J Curr Microbiol App Sci 5(6):115–124

von Arx JA (1981) The genera of fungi sporulating in pure culture. J Cramer publication, Vaduz, pp 1–424

Yuan ZQ, Zhao ZY (1992) A new species of *Broomella* and its new anamorph on *Clematis* from China. Sydowia 44:90–99

Future Perspectives and Challenges

Abstract Studies on endolichenic fungi and their secondary metabolites were started in recent past and in light of the published literature endolichenic fungi can be considered as a potential area for research. These fungi can contribute in the search for hidden fungal diversity and bio-prospection of their secondary metabolites is the proof of their significance as bio-control agents for a wide range of diseases. This chapter points out various possible queries about these intriguing creatures and their possible explanation in detail.

Various workers have assessed different lichen species for their endolichenic fungal diversity and not only found a plethora of fungi residing asymptomatically within the thallus, but also variety of secondary metabolites comprising diverse and novel structures possessing various bioactivities.

Since fungi being heterotrophic in nature are supposed to search a habitat where the possibility of finding other fungi is very low, the mere presence of endolichenic fungi within lichens, which itself is a very complex system having symbiotic association of fungi and algae, led us to think that what they are doing there and how they are surviving in such an inhospitable environment where lichen mycobiont is flourishing.

The presence of endolichenic fungal association in thalli of fossil lichen from the Lower Devonian (Honegger et al. 2013), indicates that endolichenism is a survival strategy and a necessary step in completion of life cycle of fungi. If this is so, then how these fungi acquired the endolichenic habit and evolved with lichens? Whether these fungi come in contact with lichen thallus at the time of its very synthesis or after the complete and successful establishment of symbiosis between the algal and fungal counterparts, are some other queries which need to be answered.

As most of the fungi behave as pathogens in free state and when are found living asymptomatically inside a thallus as endophyte, then do not produce any apparent disease symptoms. What is the objective and mechanism behind living asymptomatically inside lichen thallus and whether these endolichenic fungi are similar to their free counterparts present in soil, air, water, parasite on plants and humans? If they are similar to the free fungi of the same species than why they are not perform-

M. Tripathi, Y. Joshi, *Endolichenic Fungi: Present and Future Trends*,
https://doi.org/10.1007/978-981-13-7268-1_8

Table 1 Some examples of endophytes that have been shown to be latent pathogens

Host	Latent Pathogen	Disease	References
Citrus spp.	*Phomopsis citri* (Sacc.) Traverso & Spessa	Stem end rot	Wright (1998)
	Neofusicoccum ribis (Slippers, Crous & M.J. Wingf.) Crous, Slippers & A.J.L. Philips	Stem end rot	Wright (1998)
	Lasiodiplodia theobromae (Pat.) Griffon & Maubl.	Stem end rot	Wright (1998)
Vitis vinifera	*Diaporthe neoviticola* Udayanga, Crous & K.D. Hyde	Leaf lesions	Mostert et al. (2000)

ing the same functions as in free state, such as production of spores and disease symptoms, and if are different than how their nature got personalized according to the host, are some other questions which need to be answered.

It has been well known that endophytic pathogens which have co-evolved with their hosts are not highly virulent (Sieber 2007) and known to be latent pathogens (Brown et al. 1998; Photita et al. 2004; Sieber 2007; Slippers and Wingfield 2007; Table 1), and if this appears to be true in case of lichens, than whether endolichenic fungi could be the possible latent phase of the lichenicolous fungi (pathogenic to lichens)? However, the lack of endolichenic fungi in Chaetothyriales concurs with the result of Arnold et al. (2009) in that lichenicolous fungi and endolichenic fungi are often phylogenetically distinct. Besides this, the fungi isolated as endolichenic fungi from lichens have been reported previously performing various kinds of other functional roles in different environmental settings (Table 2) and their presence in the form of endolichenism could be a latent phase of the whole life cycle of pathogenic microbes.

Till date very insufficient knowledge about the diversity, ecology and biology of endolichenic fungi is available, across the globe, hence, further studies need to be carried out in a massive way to understand their life cycle and nutritional requirements in-vitro and in-vivo both. Not only this, studies on assessment of lichen diversity along altitudinal gradient have been dealt very nicely throughout the world, but diversity of endolichenic fungi along an altitudinal gradient need to be carried out.

Why some endolichenic fungi are host specific while others have broad host range, and how host specificity of one's differ from those which shows broad host ecologies? Is there any relation between host specificity of endolichenic fungi and lichen secondary metabolites? Some of the endolichenic fungi are host specific to certain lichen species while some commonly occurred in majority of lichens, so how these fungi maintain the loyalty to a particular host. The specificity factor may be genetically controlled since the development of the fungi acts in response to physical or chemical conditions which are unique to the host and act as signal to the fungus – the fungus 'recognizes' its host. The actual reason behind the specificity of

Table 2 Various functional roles played by fungal species which were also isolated as endophytes

Fungi	Functional roles
Alternaria alternata (Fr.) Keissl.	Causes leaf spot, rots and blights and other diseases in plants
Aspergillus niger Tiegh.	Causes black mould in onions and ornamental plants, and otomycosis (fungal ear infection)
Aspergillus ochraceus G. Wilh.	Pathogenic to agricultural commodities, farmed animals and marine species (Anderson et al. 1993; Wilson et al. 2002; Cui et al. 2010; Ostry et al. 2013). Consumption of this fungus by humans and animals produces chronic neurotoxic, immunosuppressive, genotoxic, carcinogenic and teratogenic effects (Ravelo et al. 2011). Asthma and lung diseases in humans (Reponen et al. 2012). Ochratoxin A, a type of mycotoxin, contaminates food and initiates apoptosis of plant cells (Wang et al. 2013b).
Fusarium oxysporum Schltdl.	Saprophyte and degrades lignin (Sutherland et al. 1983; Rodriguez et al. 1996) and complex carbohydrates (Christakopoulos et al. 1995, 1996; Snyder and Hansen 1940). Pervasive plant endophyte that can colonize plant roots (Katan 1971; Gordon et al. 1989) and may even protect plants or be the basis of disease suppression (Larkin et al. 1993; Lemanceau et al. 1993)
Neocosmospora solani ((Mart.) L. Lombard & Crous) *(= Fusarium solani)*	Causes sudden death syndrome in Soybean (Bennett et al. 2014)
Nigrospora oryzae (Berk. & Broome) Petch	Saprotroph and can be found colonizing debris of different plant species and a week parasite producing grain spots of rice, sorghum and corn (Neergaard 1977).
Penicillium citrinum Thom	Mortality for the mosquito *Culex quinquefasciatus* (Da Costa and De Oliveira 1998; Maketon et al. 2014).
Sordaria fimicola (Roberge ex Desm.) Ces. & De Not.	Fungus commonly found in the feces of herbivores.
Trichoderma viride Pers.	A biofungicide and is used for seed treatment for suppression of various diseases caused by fungal pathogens. It has been shown to provide protection against pathogens such as *Rhizoctonia*, *Pythium* and even *Armillaria*.
Xylaria hypoxylon (L.) Grev.	The compounds Xylarial A and B isolated from this species have moderate cytotoxic activity against the human hepatocellular carcinoma cell line Hep G2 (Gu and Ding 2008). The compounds Xylarone and 8,9-dehydroxylarone also have cytotoxic activity (Schüffler et al. 2007). The carbohydrate-binding protein, a lectin, with unique sugar specificity, isolated from this fungus has potent anti-tumor effects in various tumor cell lines (Liu et al. 2006).

endolichenic fungi to its host lichen is yet to be found. Petrini et al. (1990) found that endolichenic fungal colonization is not host-specific and microhabitat has no clear influence on endolichenic fungi but Tripathi et al. (2014) found that some endolichenic fungal species are common among various lichens species and some are specific to a particular lichen species. So we need more number of extensive studies on endolichenic fungi because we do not know the extent of thallus colonization and its effect on the lichen physiology.

It is a well-known fact that in lichen thallus, the algal partner fixes the atmospheric carbon and obligate fungal partner uses it and gives in return shelter to the algal partner. So the question which needs special emphasis here is to what extent endolichenic fungi share photosynthate produced by algal partner with the obligate fungal partner and though as it has been made very clear by earlier studies that one lichens thallus harbors various species of endolichenic fungi, so how all these endolichenic fungal individuals share the common food? The contribution of endolichenic fungi to the uptake of mineral nutrients and the nutrients exchange between endolichenic fungi and lichens have still not been established with compelling data, though some studies in case of fungal endophytes colonizing higher plant suggests indirect aid of nutrients by the endophytes (Harman 2000; Usuki and Narisawa 2007; Upson et al. 2009; Newsham 2011).

It has been observed in various cases that some fungal endophytes colonizing root of higher plants have positive effects on the growth of its host (Yedidia et al. 2001; Adams et al. 2007; Shoresh et al. 2010), enhanced fruit production (Andrade-Linares et al. 2011, 2013), positive regulation of proteins implicated in stress and defense responses in addition to carbohydrate metabolism and a photosynthetic augmentation (Shoresh and Harman 2008), improved seed germination and growth under abiotic and biotic stresses (Mastouri et al. 2010). Hence, likewise endophytes of higher plants, does presence or absence of endolichenic fungi have any role in the growth of lichen thallus? This objective need to be worked out in case of lichens and by doing this we can relate the pollution tolerance, great diversity, dominance in any ecosystem, host and habitat selectivity, endemism of some particular lichen species to the endolichenic fungal composition which they harbor.

There are many species of lichens which used to grow in natural ecosystem exposed to unfavorable climatic conditions, such as, very low temperatures, high UV radiations and polluted environments. Such stressful circumstances could be meliorated by mutualistic symbioses and regarding lichens whether this symbiotic partner could be the endolichenic fungi need to be answered. So far, there are no evidences to support that the endolichenic fungi could be responsible for the growth and development of lichens in harsh climatic conditions, but just for sake, if we look at what endophytes in higher plants are doing, viz. *Fusarium*, *Colletotrichum* and *Curvularia* confer plant resistance to salt, heat and drought of tomato, rice and dunegrass plants (Rodriguez et al. 2008; Redman et al. 2011) while *Trichoderma* isolates confer drought tolerance (Shukla et al. 2012), then couldn't there be a possibility that endolichenic fungi can perform the same functions in lichens which needs to be checked out in near future.

Till date various forms of symbiotic relations in natural ecosystems have been generally classified into mutualism, commensalism and parasitism on the basis of the profit which is shared by both the partners (Paszkowski 2006), but the relationship between endolichenic fungi and lichens have not been categorized in any of these said categories. The endolichenic fungi are phylogenetically diverse offering a variable host response, so consequently there is a wide range of interaction between lichen and endolichenic fungi. Positive or negative effects of these endolichenic fungi appear to be regulated not only by the genes of the major share holder "the mycobiont" and by the other obligate partner "the phycobiont" but also by other factors such as the nutritional status of endolichenic fungi and lichen, and by environmental settings during the interaction. The type of interaction is incredibly complicated to define, that whether it is mutualistic or commensalistic or parasitic without realizing the actual costs and benefits for both associates. The physiological functions of endolichenic fungi are insufficiently described. *Alternaria alternata* produces extracellular enzymes (amylase, laccase, pectinase, and protease) (Sun et al. 2011); this suggests its potential role in litter degradation, and it may be helpful in decomposing the lichen thallus after senescence. Investigating the functional roles of endolichenic fungi appears more fascinating in combination with various modern techniques, such as genomics, transcriptomics, proteomics, which may help us to answer whether the endolichenic fungi may exist in a living state and become pioneer decomposers after lichen senescence or become a pathogen consequently instigate a disease as the thallus is in unfavorable conditions and then be the indirect cause of its death.

Results of the previous studies revealed that lichen mycobiont produce various biologically active secondary metabolites and in recent past endolichenic fungi have also become a great source of secondary metabolites. Both these fungi live in close association inside the lichen thalli so is there any confrontation or synergy between metabolites produced by endolichenic fungi and lichen mycobiont?

Isolation, purification and structural identification of secondary metabolites associated with endolichenic fungi is also not an easy task and many technical challenges still exist in this process which need to be overcome in future researches to create a new horizon for the search of novel natural products for the benefit of human kind.

It is well known that endophytes which have the capacity of producing the metabolites which imitate the structure and function of metabolites synthesized by their host plants have shifted the focus of new drug production towards endophytes. The evidence of metabolites that are synthesized by both partners has raised the question of horizontal gene transfer and/or a cross talk of metabolic pathways between endophyte and plant (Stierle et al. 1993; Sirikantaramas et al. 2009; Kusari et al. 2012). Though the metabolites produced by endolichenic fungi didn't mimic the structure of metabolites produced by their host lichens but show promising potential against various microbes, hence, their functioning needs to be evaluated and if metabolites isolated from lichens and endolichenic fungi shows similar functioning then we can save tons of lichens from harvesting for the purpose of making dyes, perfumes, spices, various drugs and can use the endolichenic fungi metabolites instead.

As it is well known that secondary metabolites of some lichen species have antifungal properties (Piovano et al. 2002; Halama and van Haluwin 2004; Goel et al. 2011; Mitrovic et al. 2011), but the surprising fact is that the endolichenic fungi which are residing inside the lichen thallus seems to be unaffected by these metabolites. Since no such studies regarding the effects of lichen metabolites on the endolichenic fungi have been carried out, hence the actual cause behind this mechanism is not known and need to be answered? May be there is a system inside lichen to detect the self fungi (endolichenic fungi) from the non-self (pathogenic) as there's in the bacteria (Restriction modification) to find the self and no-self DNA molecule.

Secondary metabolites produced by endolichenic fungi have a range of pharmacological properties, including antioxidant, anticancer, antifungal, antibacterial, antiquorum sensing, antiviral, and anti-Alzheimer's properties. Majority of these bioactive properties have been performed in-vitro but in-vivo studies need to be carried out in support so that they can be used as new bioresource for therapeutic metabolites.

Since all microbial resources sustain a degree of biosynthetic flexibility in metabolite synthesis hence, the technique 'one strain, many compounds' (OSMAC) can be used to enhance the number of metabolites produced by a microbial strain. The OSMAC technique alters the culture conditions viz. change in media composition, temperature, pH, aeration, addition of enzyme inhibitors, and epigenetic modifiers, which lead to increase in number of metabolites production. For example, bioactivity-guided isolation of *Ulocladium* sp. from *Everniastrum* sp. yielded five additional novel ophiobolins when cultivated in potato dextrose broth. One of these, ophiobolin T, showed strong cytotoxic activity against hepatocellular carcinoma HepG2 cells and demonstrated moderate antibacterial activity against *Bacillus subtilis* and methicillin-resistant *Staphylococcus aureus* (Wang et al. 2013a). Thus in the development of drugs the production of culture filtrate from endolichenic fungi is documented as one of the fundamental predicament. As we know that the huge amounts of culture filtrate are required to attain the structure and a fair amount of a specific compound, various other alternative approaches, as semi-synthetic production of compounds, are needed for further experiments. Since the production of secondary metabolites by fungi is a somewhat slow process if compared to other fast growing microbes, hence, through the utilization of 'omics' techniques to examine the secondary metabolites production pathway, genome editing, and installing whole pathways into fast-growing microbes, including yeast and bacteria, the production can be achieved in a short span of time and the particular compound can be produced industrially at any time (Ro et al. 2006).

In this era of nanotechnology, where scientists are in search for nano-particles from various sources, fungi also not lag far behind. Till date many fungi viz. *Verticillium* (Mukherjee et al. 2001), *Phoma* sp. (Chen et al. 2003; Birla et al. 2009), *Phanerochaete chrysosporium* (Vigneshwaran et al. 2006), *Aspergillus niger* (Gade et al. 2008), *Aspergillus fumigatus* (Bhainsa and D'Souza 2006), *Aspergillus flavus* (Vigneshwaran et al. 2007), *Fusarium oxysporum* (Ahmad et al. 2003; Senapati et al. 2004; Duran et al. 2005, 2007), *Fusarium semitectum* (Basavaraja et al. 2008), *Fusarium acuminatum* (Ingle et al. 2008), *Fusarium solani* (Ingle et al. 2009), *Penicillium* (Sadowski et al. 2008), *Trichoderma asperellum* (Mukherjee et al. 2008), *Coriolus versicolor* (Sanghi and Verma 2009) and *Cladosporium cladosporioides* (Balaji et al. 2009), have been used to produce silver nanoparticles intracellularly or extracellularly which were of particular relevance to new and emerging technologies. For example Ingle et al. (2008) evaluated the biosynthesized silver nanoparticles from *Fusarium acuminatum* for their broad spectrum antibacterial activity on different human pathogens and reported efficient antibacterial activity of silver nanoparticles against *Staphylococcus aureus*, *Staphylococcus epidermidis*, *Salmonella typhi* and *Escherichia coli*. Another application of silver nanoparticles is the production of sterile materials. Cotton fabrics incorporated with 2% of silver nanoparticles produced by *F. oxysporum* (Duran et al. 2007), exhibited high antibacterial effect against S. *aureus* (99.9% bacterial reduction). Cai et al. (2005) reported the nanotube spearing approach, in which they used nickel embedded carbon nanotubes coated with DNA. They further reported that when the nanotubes were introduced in the cells in presence of specially oriented magnetic field, the nanotubes align with the magnetic flux lines as they are pulled towards the cells. This enables the nanotubes to spear the cells, pass through the membrane and deliver the targeted DNA. Though, the reports of production of the nanoparticles from fungi are numerous, but not a single endolichenic fungus has been used in making nanoparticles, and this needs to be worked out in near future.

In a nut shell most of the literature on endolichenic fungi is in relevance to their ability to produce novel secondary metabolites having some biological activity (Kellogg and Raja 2016). It is quite essential to know more about the biology of endolichenic fungi to appreciate their role in the establishment and preservation of the lichen symbiosis and their communications with the lichen microbiome. Hence, studies on endolichenic fungi will unbolt the entirely new prospects for research and this endless natural resource can be used in several ways for the betterment of human life.

References

Adams P, De-Leij FA, Lynch JM (2007) *Trichoderma harzianum* Rifai 129522 mediates growth promotion of Crack willow (*Salix fragilis*) saplings in both clean and metal-contaminated soil. Microb Ecol 54:306–313

Ahmad A, Mukherjee P, Senapati S (2003) Extracellular biosynthesis of silver nanoparticles using the fungus *Fusarium oxysporum*. Colloids Surf B: Biointerfaces 28(4):313–318

Anderson KH, Domsch W, Gams Traute-Heidi (1993) Compendium of soil fungi (Reprint [derAusg. London] 1980 ed) Eching: IHW-verl. ISBN 3980308383

Andrade-Linares DR, Grosch R, Franken P et al (2011) Colonization of roots of cultivated *Solanum lycopersicum* by dark septate and other ascomycetous endophytes. Mycologia 103:710–721

Andrade-Linares DR, Anja Müller A, Fakhro A et al (2013) Impact of *Piriformospora indica* on tomato. In: Varma A (ed) *Piriformospora indica*, soil biology, vol 33. Springer, Berlin, pp 107–117

Arnold AE, Miadlikowska J, Higgins K et al (2009) A phylogenetic estimation of trophic transition networks for ascomycetous fungi: are lichens cradles of symbiotrophic fungal diversification? Syst Biol 58:283–297

Balaji DS, Basavaraja S, Deshpande R et al (2009) Extracellular biosynthesis of functionalized silver nanoparticles by strains of *Cladosporium cladosporioides* fungus. Colloid Surf B: Biointerfaces 68:88–92

Basavaraja S, Balaji SD, Lagashetty A et al (2008) Extracellular bio synthesis of silver nanoparticles using the fungus *Fusarium semitectum*. Mater Res Bull 43:1164–1170

Bennett MJ, Rhetoric E, Hicks E et al (2014) The Minnesota Soybean Field Book. University of Minnesota, St Paul, MN

Bhainsa KC, D'Souza SF (2006) Extracellular biosynthesis of silver nanoparticles using the fungus *Aspergillus fumigatus*. Colloid Surf B Biointerfaces 47:160–164

Birla SS, Tiwari VV, Gade AK et al (2009) Fabrication of silver nanoparticles by *Phoma glomerata* and its combined effect against *Escherichia coli*, *Pseudomonas aeruginosa* and *Staphylococcus aureus*. Lett Appl Microbiol 1:2173–2179

Brown KB, Hyde KD, Guest DI (1998) Preliminary studies on endophytic fungal communities of *Musa acuminata* species complex in Hong Kong and Australia. Fungal Divers 1:27–51

Cai D, Mataraza JM, Qin ZH et al (2005) Highly efficient molecular delivery into mammalian cells using carbon nanotube spearing. Nat Methods 2(6):449–454

Chen JC, Lin ZH, Ma XX (2003) Evidence of the production of silver nanoparticles via pretreat ment of *Phoma* sp. 32883 with silver nitrate. Lett Appl Microbiol 37:105–108

Christakopoulos P, Kekos D, Macris BJ et al (1995) Purification and mode of action of a low molecular mass endo-1,4-B-D-glucanase from *Fusarium oxysporum*. J Biotechnol 39:85–93

Christakopoulos P, Nerinckx W, Kekos D et al (1996) Purification and characterization of two low molecular mass alkaline xylanases from *Fusarium oxysporum* F3. J Biotechnol 51:181–180

Cui CM, Li XM, Meng L et al (2010) 7-nor-ergosteroide, a pentalectone-containing norsteroid & related steroids from the marine-derived endophytic *Aspergillus ochraceous* EN-31. J Nat Prod 73(11):1780–1784

Da Costa GL, De Oliveira PC (1998) *Penicillium* species in mosquitoes from two Brazilian regions. J Basic Microbiol 38(5–6):343–347

Duran N, Marcato PD, Alves OL et al (2005) Mechanistic aspects of bio synthesis of silver nanoparticles by several *Fusarium oxysporum* strains. J Nanobiotech 3:1–8

Duran N, Marcato PD, De S et al (2007) Antibacterial effect of silver nanoparticles produced by fungal process on textile fabrics and their effluent treatment. J Biomed Nanotechnol 3:203–208

Gade AK, Bonde PP, Ingle AP et al (2008) Exploitation of *Aspergillus niger* for synthesis of silver nanoparticles. J Biobaased Mater Bioenergy 2:1–5

Goel M, Dureja P, Rani A et al (2011) Isolation, characterization and antifungal activity of major constituents of the Himalayan lichen *Parmelia reticulata* Tayl. J Agric Food Chem 59:2299–2307

Gordon TR, Okamoto D, Jacobson DJ (1989) Colonization of muskmelon and non-susceptible crops by *Fusarium oxysporum* f. sp. *melonis* and other species of *Fusarium*. Phytopathology 79:1095–1100

Gu W, Ding H (2008) Two new tetralone derivatives from the culture of *Xylaria hypoxylon* AT-028. Chinese Chem Lett 19(11):1323–1326

Halama P, van Haluwin C (2004) Antifungal activity of lichen extracts and lichenic acids. BioControl 49:95–107

Harman GE (2000) Myths and dogmas of biocontrol: Changes in perceptions derived from research on *Trichoderma harzianum* T-22. Plant Dis 84:377–393

Honegger R, Axe L, Edwards D (2013) Bacterial epibionts and endolichenic actinobacteria and fungi in the lower devonian lichen *Chlorolichenomycites salopensis*. Fungal Biol 117:512–518

Ingle A, Gade A, Pierrat S et al (2008) Mycosynthesis of silver nanoparticles using the fungus *Fusarium acuminatum* and its activity against some human pathogenic bacteria. Curr Nanosci 4:141–144

Ingle A, Gade A, Bawaskar M et al (2009) *Fusarium solani*: a novel biological agent for the extracellular synthesis of silver nanoparticles. J Nanopart Res 11:2079–2085

Katan J (1971) Symptomless carriers of the tomato *Fusarium* wilt pathogen. Phytopathology 61:1213–1217

Kellogg JJ, Raja HA (2016) Endolichenic fungi: a new source of rich bioactive secondary metabolites on the horizon. Phytochem Rev 16(2):271–293

Kusari S, Hertweck C, Spiteller M (2012) Chemical ecology of endophytic fungi: origins of secondary metabolites. Chem Biol 19(7):792–798

Larkin RP, Hopkins DL, Martin FN (1993) Effect of successive watermelon plantings on *Fusarium oxysporum* and other microorganisms in soils suppressive and conducive to fusarium wilt of watermelon. Phytopathology 83:1097–1105

Lemanceau P, Bakker PAHM, DeKogel WJ et al (1993) Antagonistic effect of nonpathogenic *Fusarium oxysporum* Fo47 and pseudobactin 358 upon pathogen *Fusarium oxysporum* f. sp. dianthi. Appl Environ Microbiol 59:74–82

Liu Q, Wang H, Ng TB (2006) First report of xylose specific lectin with potent hemagglutinating, antiproliferative and antimitogenic activities from a wild ascomycete mushroom. Biochim Biophys Acta 1760(12):1914–1919

Maketon M, Amnuaykanjanasin A, Kaysorngup A (2014) A rapid knockdown effect of *Penicillium citrinum* for control of the mosquito *Culex quinquefasciatus* in Thailand. World J Microbiol Biotech 30(2):727–736

Mastouri F, Bjorkman T, Harman GE (2010) Seed treatment with *Trichoderma harzianum* alleviates biotic, abiotic, and physiological stresses in germinating seeds and seedlings. Phytopathology 100:1213–1221

Mitrovic T, Stamenkovic S, Cvetkovic V et al (2011) Antioxidant, antimicrobial and antiproliferative activities of five lichen species. Int J Mol Sci 12:5428–5448

Mostert L, Crous PW, Petrini O (2000) Endophytic fungi associated with shoots and leaves of *Vitis vinifera*, with specific reference to the *Phomopsis viticola* complex. Sydowia 52:46–58

Mukherjee P, Ahmad A, Mandal D (2001) Fungus-mediated synthesis of silver nanoparticles and their immobilization in the mycelial matrix: a novel biological approach to nanoparticle synthesis. Nano Lett 1:515–519

Mukherjee P, Roy M, Mandal BP et al (2008) Green synthesis of highly stabilized nanocrystalline silver particles by a non-pathogenic and agriculturally important fungus *T. asperellum*. Nanotechnology 19(7):075103

Neergaard P (1977) Seed Pathology. The Macmillan Press Ltd., London

Newsham KK (2011) A meta-analysis of plant responses to dark septate root endophytes. New Phytol 10:783–793

Ostry V, Malir F, Rupich J (2013) Producers & important dietary sources of ochratoxin A & citrinin. Toxins 5(9):1574–1586

Paszkowski U (2006) A journey through signaling in arbuscular mycorrhizal symbioses. New Phytol 172(1):35–46

Petrini O, Hake U, Dreyfuss M (1990) An analysis of fungal communities isolated from fruticose lichens. Mycologia 82:444–451

Photita W, Lumyong S, Lumyong P et al (2004) Are some endophytes of *Musa acuminata* latent pathogens? Fungal Divers 16:131–140

Piovano M, Garbarino JA, Giannini FA et al (2002) Evaluation of antifungal and antibacterial activities of aromatic metabolites from lichens. Bol Soc Chil Quım 47:235–240

Ravelo AA, Rubio AC, Gutierrez FAJ et al (2011) Ochratoxin A in foods for Human consumption: review. Nutr Hosp 26(6):1215–1226

Redman RS, Kim YO, Woodward CJDA et al (2011) Increased fitness of rice plants to abiotic stress via habitat adapted symbiosis: a strategy for mitigating impacts of climate change. PLoS One 6(7):e14823. https://doi.org/10.1371/journal.pone.0014823

Reponen T, Lockey J, Bernstein DI et al (2012) Infant origins of childhood asthma associated with specific moulds. The J Allergy Clin Immunol 130(3):639–644

Ro DK, Paradise EM, Ouellet M et al (2006) Production of the antimalarial drug precursor artemisinic acid in engineered yeast. Nature 440:940–943

Rodriguez A, Perestelo F, Carnicero A et al (1996) Degradation of Natural lignins lignocellulosic substrates by soil inhabiting fungi imperfecti. FEMS Microbial Ecol 21:213–219

Rodriguez RJ, Henson J, Van Volkenburgh E et al (2008) Stress tolerance in plants via habitat-adapted symbiosis. ISME J 2(4):404–416

Sadowski Z, Maliszewska IH, Grochowalska B et al (2008) Synthesis of silver nanoparticles using microorganisms. Mater Sci Pol 26:419–424

Sanghi R, Verma P (2009) A facile green extracellular biosynthesis of CdS nanoparticles by immo bilized fungus. Chem Eng J 155:886–891

Schüffler A, Sterner O, Anke H (2007) Cytotoxic α-Pyrones from *Xylaria hypoxylon*. Z Naturforsch 62(c):169–172
Senapati S, Manda D, Ahmad A (2004) Fungus mediated synthesis of silver nanoparticles: a novel biological approach. Indian J Phys 78(1):101–105
Shoresh M, Harman GE (2008) The molecular basis of shoot responses of maize seedlings to *Trichoderma harzianum* T22 inoculation of the root: a proteomic approach. Plant Physiol 147:2147–2163
Shoresh M, Harman GE, Mastouri F (2010) Induced systemic resistance and plant responses to fungal biocontrol agents. Annu Rev Phytopathol 48:1–23, 21
Shukla N, Awasthi RP, Rawat L et al (2012) Biochemical and physiological responses of rice (*Oryza sativa* L.) as influenced by *Trichoderma harzianum* under drought stress. Plant Physiol Biochem 54:78–88
Sieber T (2007) Endophytic fungi in forest trees: are they mutualists? Fungal Biol Rev 21:75–89
Sirikantaramas S, Yamazaki M, Saito K (2009) A survival strategy: the coevolution of the camptothecin biosynthetic pathway and self-resistance mechanism. Phytochemistry 70:1894–1898
Slippers B, Wingfield MJ (2007) Botryosphaeriaceae as endophytes and latent pathogens of woody plants: diversity, ecology and impact. Fungal Biol Rev 21:90–106
Snyder WC, Hansen HN (1940) The species concept in *Fusarium*. Amer J Bot 27:64–67
Stierle A, Strobel G, Stierle D (1993) Taxol and taxane production by *Taxomyces andreanae*, an endophytic fungus of Pacific yew. Science 260:214–216
Sun X, Guo LD, Hyde KD (2011) Community composition of endophytic fungi in *Acer truncatum* and their role in decomposition. Fungal Divers 47:85–95
Sutherland JB, Pometto ALIII, Crawford DL (1983) Lignocellulose degradation by *Fusarium* species. Can J Bot 61:1194–1198
Tripathi M, Joshi Y, Gupta RC (2014) Assessment of endolichenic fungal diversity in some forests of Kumaun Himalaya. Curr Sci 107(5):745–748
Upson R, Read DJ, Newsham KK (2009) Nitrogen form influences the response of *Deschampsia antarctica* to dark septate root endophytes. Mycorrhiza 20(1):1–11
Usuki F, Narisawa K (2007) A mutualistic symbiosis between a dark septate endophytic fungus, *Heteroconium chaetospira*, and a nonmycorrhizal plant, Chinese cabbage. Mycologia 99:175–184
Vigneshwaran N, Kathe AA, Varadarajan PV et al (2006) Biomimetics of silver nanoparticles by white rot fungus, *Phanerochaete chrysosporium*. Colloids Surf B: Biointerfaces 53:55–59
Vigneshwaran N, Ashtaputre NM, Varadarajan PV et al (2007) Biological synthesis of silver nanoparticles using the fungus *Aspergillus flavus*. Mater Lett 61:1413–1418
Wang QX, Bao L, Yang XL et al (2013a) Ophiobolins P-T, five new cytotoxic and antibacterial sesterterpenes from the endolichenic fungus *Ulocladium* sp. Fitoterapia 90:220–227
Wang Y, Hao J, Zhao W et al (2013b) Comparative proteomics & physiological characterization of *Arabidopsis thaliana* seedlings in responses to ochratoxin A. Plant Mol Biol 82(4–5):321–337
Wilson DM, Mubatanhema W, Jurjevic Z (2002) Biology and ecology of mycotoxigenic *Aspergillus* species as related to economic and health concerns. Adv Experiment Med Biol 504:3–17
Wright JD (1998) The role of endophytes in Citrus stem end rots. Ph.D. Thesis, The University of Hong Kong, Hong Kong
Yedidia I, Shrivasta AK, Kapulnik Y et al (2001) Effect of *Trichoderma harzianum* on microelement concentration and increased growth of cucumber plants. Plant Soil 235:235–242

MIX
Papier aus verantwortungsvollen Quellen
Paper from responsible sources
FSC® C105338

If you have any concerns about our products,
you can contact us on
ProductSafety@springernature.com

In case Publisher is established outside the EU,
the EU authorized representative is:
Springer Nature Customer Service Center GmbH
Europaplatz 3, 69115 Heidelberg, Germany

Printed by Libri Plureos GmbH
in Hamburg, Germany